R. Morison

Narrative of the Shipwreck of the Antelope East-India Pacquet,

on the Pelew Islands, situated on the western part of the Pacific Ocean, in August,

1783

R. Morison

Narrative of the Shipwreck of the Antelope East-India Pacquet,
on the Pelew Islands, situated on the western part of the Pacific Ocean, in August, 1783

ISBN/EAN: 9783337316204

Printed in Europe, USA, Canada, Australia, Japan

Cover: Foto ©berggeist007 / pixelio.de

More available books at **www.hansebooks.com**

PREFACE.

NO subject can be more interesting to man, than the history of man; and no method can be more proper for investigating this subject, than examining the different appearances which he assumes in different quarters of the globe. History does not offer an example of such disinterested efforts, towards the enlargement of human knowledge, as have been made by the British nation, since the accession of his present Majesty to the throne—the noble and hazardous enterprizes undertaken and executed under his auspices, will remain to succeeding ages, a monument of the zeal and patronage of GEORGE THE THIRD.

If to bring to view a part of our Brethren of Mankind hitherto unknown—a race of men that do honour to the name of Manhood, be a subject generally interesting, the following pages must ensure a favourable reception. The Public have been often amused with uninteresting histories of voyages and shipwrecks; and fiction and fancy have been tortured to throw into the dish something palateable; it may therefore be necessary to mention a few facts as to the authenticity of a narrative, which arrests the attention with incidents, seldom surpased in the airy visions of romance.

The

PREFACE.

The Antelope was not fitted out for the purpoſe of adventure or diſcovery; it was a pacquet commanded by Captain Henry Wilſon, and manned by a number of hands the greater part of whom are now in Britain, and can vouch for the reality of every circumſtance. They were wrecked on this almoſt unknown coaſt, and after ſuffering a ſeries of unparallelled hardſhips, were reſtored to their country, their home and their friends, by the benevolence of this new race of men.

It is ſomewhat remarkable that although the iſlands which gave birth to ſome of the moſt important ſcenes in this work, lie at no very great diſtance from the common track to China, yet it does not appear that any Europeans ever landed on them. They were not however totally unknown. In the Lettres *edifientes et curieufes*, we find an account of this Archipelago, of which the Palos or Pelew Iflands conftitute the fifth divifion; the other four confifting of the Iflands which are now known by the name of the *New Carolines*. Le Pere Cantora tells us, that being fhipwrecked on one of the Caroline Iflands, he had ufed every method to get information concerning the reft; and that he was informed, " that the people of the Pelew Iflands were inhuman and favage; that both men and women were entirely naked, and fed upon human flefh; that the inhabitants of the

Carolines

PREFACE.

Carolines looked on them with horror as the enemies of mankind, and with whom they held it dangerous to have any intercourfe." From this and any other information that can be procured, it appears that for a long feries of years, the inhabitants of the Pelew Iflands, have been feparated from the reft of mankind, even thofe moft contiguous to themfelves. Their ignorance of the exiftence of white people, abundantly evinces their being total ftrangers to Europe at any rate.

The name given by the Spaniards to thefe Iflands is the Palos Iflands; which indeed is the name by which all the Caroline Iflands formerly went; probably owing to the number of tall palm trees

trees with which they are covered, having the appearance of masts of ships at a distance, the Spanish word *palos* signifies a mast.

If therefore this publication tends to introduce a *new people* to the reader, who have, uninformed and untaught as they are, brought their manners to a great degree of civilization; the politician, the moralist, and the friend of humanity, will not consider this an useless work, more especially to those who have not access to peruse Mr Keate's very complete publication on the subject.

CONTENTS.

CONTENTS.

CHAP. I.

The Antelope Sails from Macao—List of the Crew—Distress—Struck on a Rock—Landing—Appearance of Natives—Interview and Transactions between them—Particulars concerning Raa Kook, Arra Kooker, &c. Page 1

CHAP. II.

The King Visits them—Reception and Behaviour—Examines every thing, and is pleased—Farther Proceedings. 43

CHAP. III.

Captain Wilson Visits Pelew—Manners of the Natives—Ten Men sent with them to Battle—Battle of Artingall—Raa Kook Visits Oroolong for more men—Death and Funeral of the King's Son. 69

CHAP. IV.

Disaster of the Vessel—Account of the Third Engagement—Captain Wilson's Visit to Rupacks

Rupacks—*Abba Thulle Visits the English with his Wife and Daughter, &c.*
Page 116

CHAP. V.

Expedition to Pelelew—Visit from Abba Thulle—Launching of the Schooner—Presents to the King—Preparations for Departure—Captain Wilson made a Rupack—Blanchard's Determination to Remain—Proposal for Lee Boo Returning with the English—Other Proceedings previous to Sailing. 150

CHAP. VI.

General Description of the Islands—Productions—Natives—Dress—Dispositions—Manners—Religion—Marriages—Customs—General Character—Government—Precedency, &c. 205

CHAP. VII.

Passage to Macao—Proceed to Canton and Embark for England—Anecdotes of Lee Boo—His Distress and Death. 239

NARRATIVE

NARRATIVE
OF THE
SHIPWRECK
OF THE
ANTELOPE.

CHAP. I.

The Antelope sails from Macao—List of the Crew—Distress—Struck on a Rock—Landing—Appearance of Natives—Interview and Transactions between them—Particulars concerning Raa Cook, Arra Kooker, &c.

CAPTAIN Henry Wilson, of the Antelope Packet, in the service of the British East India Company, about 300 tons burthen, sailed from Macao, in China, on her passage homeward, on Sunday the 20th, July 1783.

As in the sequel of this narrative, there will be occasion to mention the names of the ship's company at times, the following list of them may be useful.

Names.	Stations.
Henry Wilson,	Commander.
Phillip Benger, *since dead*,	Chief Mate.
Peter Barker,	Second Mate.
John Cummin,	Third Mate.
John Sharp,	Surgeon.
Arthur William Devis,	Passenger.
John Blanch,	Gunner.
William Harvey,	Boatswain.
John Polkinghorn,	Carpenter.
John Meale,	Cooper and Steward.
Richard Jenkins,	Carpenter's Mate.
James Swift,	Cook.
Richard Sharp,	Midshipman.
Henry Wilson, Junr.	Midshipman, *son to the Captain*.
John Wedgebrough,	Midshipman.
Robert White,	Midshipman.
Albert Pierson,	Quarter Master.
Godfrey Minks, *since dead*,	Quarter Master.
Thomas Dutton,	Captain's Steward.
Thomas Rose, a Portuguese,	Linguist.

And, Matthias Wilson, the Captain's brother, Thomas Wilson,

Wilson, *(since dead)* Dedrick Windler, Zachariah Allen, John Cooper, James Bluitt, Thomas Castles, William Roberts, Nicholas Tyacke, William Stewart, Madan Blanchard, Thomas Whitfield, William Cobbledick, and James Duncan, Seamen. Besides these Captain Wilson was allowed sixteen Chinese, to keep the ship's complement of hands complete.

On Monday the 21st they got clear to sea, having discharged the Pilot, and taken leave of several gentlemen who accompanied them a few leagues. From the 23d July, to the 8th of August, they had very stormy unsettled weather, insomuch that their foretopmast sprung, and all their live cattle died. On the 9th the weather became more moderate; so that opening their ports they dried the ship, examined their stores and provisions, and proceeded cheerfully on their voyage, flattering themselves their distress and danger were now fairly past; little judging, that the hard misfortunes they were

were about to undergo, were so quickly to overtake them.

Early on Sunday morning, the 10th of October, a strong breeze sprung up, attended with much rain, thunder and lightning. Captain Wilson had gone to bed about twelve, and Mr Benger the chief mate, commanded on deck. While the seamen were busied in reefing the sails, the man on watch exclaimed, *Breakers,* which he had scarce pronounced when the ship struck. It is not easy to express the consternation which ensued; all who were in bed below, were immediately on deck, enquiring the occasion of the noise and confusion: too soon they learned their dismal situation; in less than an hour she bulged, and filled with water up to the lower deck hatchways. During this scene of horror and dismay, the seamen eagerly besought the Captain

to

to direct them, and his orders would be implicitly obeyed.

Captain Wilson's first orders were, to secure the gun-powder and small arms, and to get on deck the bread, and such other provisions as were liable to be spoiled by the water, and cover them with tarpaulins, &c. to keep them from the rain. As the ship took a heel in filling, there was some reason to fear she might overset; to prevent which, they cut away the mizen-mast, the main and fore top-masts, and lowered the fore and main-yards, to ease her. The boats were then hoisted out, and filled with provisions; a compass, and some small arms, with ammunition, and two men, being put into each with directions to keep them under the lee of the ship, and to be ready to receive their ship-mates, in case the vessel should part by the violence of the wind and waves,

waves, as it then blew an exceeding strong gale.

Every thing being now done that prudence could dictate in so trying and distressful a situation, the officers and people assembled on the quarter deck, that part being highest out of the water, and best sheltered from the rain and sea by the quarter-boards; and waited for day-light, in hopes of seeing land, for as yet they had not been able to discern any. During this dreadful interval, the anxiety and horror of which is much easier to be imagined than described, Captain Wilson endeavoured to revive the drooping spirits of his crew, by reminding them, that shipwreck was a misfortune to which navigators were always liable; and that although theirs was rendered more difficult and distressing by its happening in an unknown and unfrequented sea,
yet

yet he wifhed to remind them that this confideration fhould only roufe them to greater activity, in endeavouring to extricate themfelves: and, above all, he begged leave to imprefs on their minds this circumftance, that whenever misfortunes, fuch as theirs, had happened, they had generally been rendered much more dreadful than they would otherwife have been, by the defpair of the crew, and by their difagreement among themfelves. To prevent which, he moft earneftly requefted each of them, feparately, not to tafte any fpirituous liquor, on any account whatever; and he had the fatisfaction to find a ready confent given to this moft important advice.

We have been the more circumftantial in our account of this part of their tranfactions, becaufe we think it difplays, in a moft remarkable manner,

the

the presence of mind which was preserved, and the prudence that was exerted by Captain Wilson in one of the most trying situations to which human nature can be exposed. It shews also, in the most inequivocal manner, the temper and disposition of his officers, and the whole crew, and pronounces their eulogium with ten thousand times the force of any words that could be used.

As they were almost worn out by the excessive labour they had undergone, two glasses of wine and some biscuit were given to every man aboard, and they waited for day-break with no little impatience, in hope of discovering land. Meantime they endeavoured to support each others spirits as much as possible, and by the Captain's direction, put on as many clothes as possible to carry with them, in the event of getting safe

safe from the wreck; and let it not be forgotten, among many other remarkable instances that occurred, in the course of this voyage, to the honour of this crew, that the utmost cordiality prevailed among them; none attempted in the hour of confusion, to touch his neighbour's property, nor to taste the *forbidden* spirits.

The dawn discovered to their view a small island, at the distance of about three or four leagues to the southward; and as the day-light increased, they saw more islands to the eastward. They now began to feel apprehensions on account of the natives, to whose dispositions they were perfect strangers: however, after manning the boats, and loading them in the best manner they were able for the general good, they were dispatched to the small island, under the direction of Mr Benger, the chief

chief mate, who was earneftly requeft-ed to eftablifh, if poffible, a friendly intercourfe with the natives, if they found any, and carefully to avoid all difagreement with them, unlefs reduced to it by the moft urgent neceffity. As foon as the boats were gone, thofe who were left in the fhip began to get the booms over board, and to make a raft for their fecurity, if the fhip fhould go to pieces, which was hourly expected: at the fame time they were under the moft painful apprehenfions for the fafety of the boats, on which all depended; not only on account of the natives, but with regard to the weather alfo, as it continued to blow very hard. But in the afternoon they were relieved from their fears on this head, by the return of the boats, with the welcome news of their having landed the ftores in fafety, and left five men to take care

of

of them; and that there was no appearance of inhabitants being on the island where they landed: that they had found a secure harbour, well sheltered from the weather, and also some fresh water. This good account revived them, and they proceeded in completing their raft with fresh vigour, having got another glass of wine with biscuit. A very distressing accident however, happened this day; the mizen-mast being found near the ship's stern, and part of the rigging entangled with the mizen chains, Godfrey Minks was employed to clear it, and whilst he was doing it, unluckily slipped overboard: the boats were immediately sent to his assistance, but without effect.

Having finished the raft, they loaded it together with the jolly boat and pinnace, with as many stores and provisions as they could bear, consistently with

with the safety of the people who were to be in them. And as the day was advancing, the Captain summoned all the people aboard; indeed, so busily were they employed in bringing as much as possible with them, it cost some pains to get them all collected. Their feelings on quitting the Antelope, going they knew not whither, were of the most distressing nature. The stoutest of the hands were put aboard the pinnace, which took the raft in tow and moved slowly on, till they had cleared the reef; while the jolly boat, which was of little service to the raft, proceeded alone to the shore, and joined their companions that had been left in the morning. They found a tent ready for their reception, and a spot of ground cleared for the stores &c.

The situation of those aboard the pinnace and the raft, till they cleared the

the reef, was terrible indeed. The great fwelling of the fea was fuch, that they repeatedly loft fight of each other, and thofe on the raft were obliged to tie themfelves to the planks with ropes to prevent their being wafhed off; whilft the horror of the fcene was increafed by the fcreams of the Chinefe, who were not accuftomed to the perils of the deep.

When they had fairly cleared the reef, they got into deep fmooth water, in the channel running between the reef and the iflands ; but on approaching the land, they found a very ftrong current which drove them confiderably to leeward. They foon found they could not refift its impetuofity, and therefore having brought the raft to a grapnel, all the hands got aboard the pinnace, to relieve the rowers. Mean while, the cargo of the jolly boat being unladen,

unladen, Captain Wilson, was returning in her, to assist those aboard the pinnace. The night was by this time dark, and the Captain overhearing them at a distance, hailed them. Those aboard the pinnace, overjoyed at the near prospect of relief, returned the haloo, in a manner so unusual, that Captain Wilson immediately concluded they were natives. He was the readier to form this idea, as he had just learned from those on shore, that from various circumstances they had reason to conclude, there had been natives on that spot very lately; he therefore retreated to shore with the utmost precipitation. Happily, however, they were soon relieved by the arrival of the pinnace, when all the company shook hands together, (need it be added,) with great cordiality. They supped on cheese, biscuit and water, and having lighted a match

match by the difcharge of a piftol, they kindled a fire in the cove, by which they dried their clothes and warmed themfelves. The night proved very uncomfortable as the weather was exceedingly tempeftuous, while the fear of the fhip going to pieces, before they could fave fuch things as they needed, not a little heightened their diftrefs. Left they fhould be furprifed by the natives, they fet a watch and flept on the ground by turns.

Next forenoon, (Monday the 11th) proved very ftormy, they attempted to bring off the raft in vain, and were obliged to leave it, bringing with them the fails and remainder of the provifions.

In the afternoon, the weather was more moderate and the boats were difpatched to the fhip to bring off what they could; while thofe on fhore were employed in brushing up the small arms.

The evening set in very squally, and as the boats did not return from the ship till about ten o'clock, those on shore were not a little alarmed about them; nor were they much easier, when on their return they learned, that the vessel was in such a situation, as made it exceedingly probable she could not hold together till morning.—When we consider their situation on this information, it must be granted, that the vicissitudes of human life, have seldom produced a coincidence of circumstances more peculiarly distressing. The only hope they had, of yet floating and repairing the vessel, so as to return to China, now to all appearance, impracticable—ignorant where they were, or among whom—separated not only from wives, children and home, but from all

all mankind, except perhaps a race of savages, as they naturally suppoſed—without any proſpect of relief—and at the ſame time ſhivering under a ſtorm ſtill more tempeſtuous than the former night, altogether brings to view a ſituation, which none can think of, even at this diſtance of time, without commiſerating.

In the morning it blew exceedingly ſtrong, ſo that the boats could not go off to the wreck: the men therefore employed themſelves in drying their proviſions, and forming better tents, from the materials which they had brought from the ſhip the day before. About eight o'clock in the morning, the people being employed as above, and in clearing the ground from the wood which was behind the tents, Captain Wilſon with Tom Roſe, a Malay, whom they had taken on board at Macao,

Macao, being on the beach, collecting the fresh water which dropped from the rocks, saw two canoes, with men in them, coming round the point into the bay. This gave such alarm, that the people all ran to their arms; however, as there were but few of the natives, Captain Wilson desired them to keep out of sight, until they should perceive what reception he met with, but to be prepared for the worst. They soon perceived that the natives had seen the Captain and Tom Rose, for they conversed together, and kept their eyes stedfastly fixed on that part of the shore where the English were. The natives advanced very cautiously toward them, and when they came near enough to be heard, the Captain directed Rose to speak to them in his own language; which they at first did not seem to understand; but they stopped their canoes,

noes, and foon after one of them afked, in the Malay tongue, who our people were, and whether they were friends or enemies? Rofe was directed to reply, that they were Englifhmen, who had loft their fhip on the reef, but had faved their lives, and were friends. On this they feemed to confer together for a fhort time, and then ftepped out of the canoes into the water, and went toward the fhore. Captain Wilfon inftantly waded into the water to meet them, and embracing them in the moft friendly manner, led them to the fhore, and prefented them to his officers, and unfortunate companions. They were eight in number, two of whom, it was afterwards known, were brothers to the Rupack, or King, of the neighbouring iflands, and one was a Malay, who had been fhipwrecked in a veffel belonging to a Chinefe, refident on the ifland of Ternate,

Ternate, one of the same group of islands; he had been kindly treated by the King, who, he said, was a good man; and that his people also were courteous. He told them farther, that a canoe having been out fishing, had seen the ship's mast; and that the King, being informed of it, sent off these two canoes at four o'clock that morning, to see what was become of the people who had belonged to her; and they knowing of the harbour which the Englishmen were in, had come directly thither.

Being about breakfast hour, Captain Wilson, Tom Rose and only a few others breakfasted with them; and in the course of their short conversation, a wish was hinted to be informed, by what means the *Malay* they had brought with them, had reached their islands. The *Malay*, who could indistinctly hammer out a few sentences both of

Dutch

Dutch and English, informed them, that he had formerly commanded a Chinese trading vessel, and about ten months since, on a voyage to Amboyna, had been cast away on a neighbouring island, from whence he had come to Pelew; this account they afterwards found reason to believe was not just. The Malay also mentioned, that one of the Pelew fishing canoes, had observed the wreck, and in consequence, these two canoes had been dispatched to succour the mariners if found. They disliked tea, but relished the biscuits very much; and in a short time grew very familiar and happy with us. After breakfast Captain Wilson introduced them to several of his officers, and acquainted them with our mode of welcoming, by shaking hands, a custom which they never afterwards omitted on meeting any of the English.

The

The natives were of a moderate fize, but admirably proportioned and very mufcular; their hair was long and black, rolled up in a peculiarly neat manner clofe to their heads; except the younger of the King's two fons, none of them had beards; they in general plucked out the hairs by the roots; and it was remarkable, that when they afterwards examined the perfons of the Englifhmen, they difcovered not a little furprife that they could fuffer hair to grow on their breafts. They were perfectly naked, and their fkin of a deep copper colour; only their thighs appeared much darker, from being tatooed very clofely. They ufed cocoa-oil, rubbing it on their fkins, which gave them a fhining appearance and very foft. The chief carried in his hand, a bafket of beetle-nut, and a very neat bamboo, which contained what they called their

chinam;

chinam; this is coral burnt to a lime, with which they fprinkle the leaf of the beetle-nut and then chew it; this makes their faliva red, which appearing betwixt their jet black teeth, occafions a very difagreeable contraft. They were conducted round the cove, and to the great furprife of the Englifh, walked on broken rock, fhells and thorny plants with the greateft eafe. They were now enabled to fupport a mutual converfation, by means of the *Malay-man,* on the part of the natives, and Tom Rofe on that of the Englifh, and thus had an opportunity of examining one another as to the different appearances which occafioned mutual furprife. From this firft interval, as well as what happened afterwards, it was evident, the natives had never before feen a white man, and were ignorant of the exiftence of

any

any such; the natural surprise at seeing them may therefore be conceived. The appearance of cloaths was quite new; indeed, they were at first at a loss to determine, whether the man and his dress were not of the same substance. One of their ideas was exceedingly natural, on observing the white skin intermixed with the blue veins, they supposed the latter to be the effects of tatooing. But nothing afforded them greater surprise than the sight of two dogs belonging to the ship, which immediately on their approach, set up a loud bark to the great delight of the natives, who answered them in a shout almost as violent; in these animals, they took great delight, as, except a few grey rats, there are no quadrupeds on the island. Captain Wilson was exceedingly anxious to keep them ignorant of the nature and use of fire-arms; but one

one of them accidentally picking up a small leaden bullet, surprised at its weight, examined the Malay about it, who requested one of our musquets, to explain it by, to him. They seemed very desirous that one of the English should go with them in their canoes to their King, that he might see what sort of people they were. Every one agreed that it would be right for some person to go; but as difficulties arose concerning who the person should be, the Captain requested his brother, Mr Matthias Wilson, who readily consented; and about noon one of the canoes left the harbour, having Mr Wilson with them: the other canoe, with four persons, among whom was Raa Kook, the elder of the King's brothers, and who was also General of his armies, remained with our people, of their own accord, until the canoe return-
turned

ed with Mr Wilson. The Captain directed his brother to acquaint the King who they were; to relate to him, as well as he could, the nature of their misfortune; to solicit his friendship and protection, and permission to build a vessel to carry them back to their own country. He also sent a present by him to the King, of a small remnant of blue broad cloath, a canister of tea, another of sugar-candy, and a jar of rusk: the last article was added at the particular request of the King's two brothers.

Those who remained behind, observing that our people had much trouble to procure the fresh water which they had occasion for, conducted them across a narrow part of the island, where it was more plentiful; and the road to it being rugged and difficult, they carried Mr Sharp, a youth of about fifteen years of age, who was sent on this duty, ty, over the moſt difficult parts, in their arms; and were very careful alſo in aſſiſting the men, in thoſe places, who brought two jars of water from the well.

The weather remained as bad as ever all the next night; but grew better in the morning; and about 10 o'clock one of the boats went to the wreck. When they got there, they found that ſome of the natives had been on board; and that they had carried off ſome iron-work, and other things; and in particular, that they had rummaged the medicine-cheſt, and taſted ſeveral of the medicines, which being probably not very palatable, they had thrown out the contents, and carried off the bottles. This circumſtance was made known to Raa Kook by the Captain, not ſo much by way of complaint, as to expreſs his uneaſineſs for the conſequences which might arise to the natives from their having tasted, or perhaps drank, such a variety of medicines. Raa Kook's countenance fully described the indignation he felt at the treacherous behaviour of his countrymen; desiring that if they caught any of them attempting again to plunder the vessel, they would kill them, and he would justify the English to his brother for having done so: and he begged that Captain Wilson would entertain no uneasiness whatever on account of what the plunderers might suffer, because it would be entirely owing to their own misconduct.

The same evening, Captain Wilson made a proposition to his officers, the boldness and wisdom of which, and the resolution manifested in its execution, reflect the highest honour on him and them, while the unanimity and cheerfulness

cheerfulness displayed by the crew in acceding to it, equally redound to their credit. Every one who knows any thing of seamen, knows that their grog " is the solace and joy of their lives;"—that their grog softens all their hardships, and causes even the horrors of war to pass by them unfelt;—that there is no undertaking so dangerous, or so desperate, that they will not most readily attempt for it, nor scarcely a comfort in life that they will not forego rather than relinquish it. Yet Captain Wilson finding them rather noisy when they returned from the wreck, (owing to a little strong liquor which the officer who was then on duty had given them, and which coming after long toil, and on an empty stomach, had operated powerfully, rather from these circumstances than from the quantity which they had drank), it alarmed him so much, that he submitted to his officers the propriety of staving (with the consent of the people) every cask of liquor which was in the ship. He knew it was too bold a step to be taken without their consent, and was fully aware of the difficulty of obtaining it; but he trusted to the regard which they had for him, and his influence over them; and he intended to make the people themselves the executioners of his purpose, while they were yet warm with the project. He had the satisfaction to find that his officers immediately acquiesced,—and the next morning he called all the crew together, and told them he had something to propose, in which their future welfare, nay, perhaps, their preservation, was most materially involved. He then submitted to their judgment the measure on which he and his officers had deliberated

ted the evening before; urged the propriety of it in the moſt forcible terms, as a ſtep which would beſt authoriſe the hope of deliverance from their preſent ſituation, and of ſeeing once more their own country, and thoſe who were dear to them;——and he added, that however reluctantly they might yield to the propoſition, yet they could not but be ſatisfied that the underſtanding of every individual amongſt them muſt, on reflection, perceive that it was a meaſure abſolutely neceſſary to be adopted. The moment he concluded, all the ſailors, with the utmoſt unanimity, and to their laſting honour as men, replied, that however they might ſuffer from being deprived of their accuſtomed recruit of liquor, yet being ſenſible, that having eaſy acceſs to it, they might not at all times uſe it with diſcretion, they gave their full aſſent to the Captain's propoſal;

proposal; and added, that they were ready to go directly to the ship, and stave every cask of liquor on board. This they most conscienciously performed; and so scrupulously did they execute their trust, that there was not a single man among them who would take even a farewel glass of his beloved cordial.

During the absence of Matthias Wilson, they had an opportunity of getting more intimately acquainted with Raa Kook, whom they found a most amiable character indeed. Observing a piece of polished bone around his wrist, they took occasion to enquire into the meaning of it. He informed them, it was a mark of great distinction, conferred only on the Blood Royal, and principal officers of state; and that he enjoyed it as being the King's brother and commander in chief of the forces both by sea and land. Raa Kook's friendship

friendship was therefore cultivated with all imaginable assiduity, and he in return shewed himself attached to them by a most attentive politeness; he imitated them in all their actions, and on every occasion shewed them how high an opinion he had formed of them. The *Malay* on his first arrival had requested the use of some cloathing which was readily granted, and an uniform coat with trowsers were at same time given to Raa Kook, who put them on, but soon wearied of them, as he found them cumbersome. He examined into the most minute action, and was at no little pains to learn from the cook, the method of blowing up the fire with a bellows.

In the morning of the 24th, two canoes arrived, in which were Arra Kooker, the King's other brother, and one of the King's sons. They informed Capt. W.

W. that his brother was on his way back; but that the canoe in which he was, could not make so much speed against the wind as theirs, which occasioned the delay. The King by their means, offered them a hearty welcome to his territories, and assured them of his friendship and protection; he also desired them to build a vessel in any part of the island they inclined, and that he and his subjects would willingly afford them every assistance in their power. Raa Kook then took his nephew and introduced him particularly to the Captain and his officers, and conducted him round the cove, explained every thing agreeable to the information he himself had just received, and seemed mightily pleased with his friend's astonishment. This young man was very well made, but had a slit in his nose, probably the consequence of a wound in battle.

In

In the forenoon two boats were difpatched to the wreck. They found a number of the natives in about 20 canoes bufied in examining the veffel; thefe Raa Kook foon difpatched, and on this as well as every occafion did every thing in his power to convince the Englifh of his protection and friendfhip.

Meantime the people were highly entertained with Arra Kooker, who proved to be a moft facetious entertaining man; poffeffing uncommon talents for mimickry and humour; he defcribed by many diverting figns the terror of Matthias Wilfon while at Pelew; indeed he had been under very great apprehenfion; but they were all revived with his appearance, and the account he gave them of his embaffy, in nearly the following words:

On the approach of the canoe in which

which I went to the ifland where the King lives, a vaft concourfe of the natives ran out of their houfes to fee me come on fhore. The King's brother took me by the hand, and led me up to the town, where a mat was fpread for me, on a fquare pavement, and I was directed to fit down on it. In a little time the King appeared, and being pointed out to me by his brother, I rofe and made my obeifance after the manner of Eaftern nations, by lifting my hands to my head, and bending my body forward; but he did not feem to pay any attention to it. I then offered him the prefents which my brother had fent by me, and he received them in a very gracious manner. His brother now talked a great deal to him, the purport of which, as I conceived, was to acquaint him with our difafter, and the number of us; after which the King

ate

ate some of the sugar-candy, seemed to relish it, and distributed a little of it to several of his chiefs, and then directed all the things to be carried to his own house. This being done, he ordered refreshments to be brought for me: the first consisted of a cocoa-nut shell of warm water, sweetened with molasses: after tasting it, he directed a little boy, who was near him, to climb a cocoa-nut tree, and gather some fresh nuts; he cleared one of them from the husk, and after tasting the milk of it, bad the boy present it to me, making signs for me to drink the milk, and then return the nut to him; and when I had done so, he broke the shell in two, ate a little of the meat, and then returned it to me to eat the rest. A great crowd of the natives had by this time surrounded me, who were curious and eager to examine my clothes

and person; but as it began to be dark, the King, his brother, myself, and several others, retired into a large house, where supper was brought in, consisting of yams boiled whole, and others boiled and beaten together, as we sometimes do potatoes; there were likewise some shell fish, but I could not determine what they were. After supper I was conducted to another house, at some distance from the first, by a female. Here I found at least forty or fifty men and women; and signs were made for me to sit or lie down on a mat, which seemed spread on purpose for me to sleep on: and after all the company had satisfied themselves with viewing me, they went to sleep, and I laid myself down on the mat, and rested my head on a log, which these people use as a pillow, and drew another mat, which also seemed laid for the
purpose,

purpose, over me. I was unable even to slumber, but lay perfectly still; and some considerable time after, when all was quiet, about eight men arose, and made two great fires at each end of the house, which was not divided by partitions, but formed one large habitation. This operation of theirs, I confess alarmed me very much indeed! I thought of nothing less, than that they were going to roast me, and that they had only laid themselves down, that I might drop asleep, for them to dispatch me in that situation. However, as there was no possibility of escaping the mischief, if any were intended, I collected all my fortitude, and recommended myself to the Supreme Disposer of all events. I lay still, expecting every moment to meet my fate; but, to my great surprise, after sitting a while to warm themselves, I perceived that they all retired again to their mats, and ſtirred no more till day-light. I then got up, and walked about, ſurrounded by great numbers of men, women, and children; and, in a little time, was joined by the King's brother, who took me to ſeveral houſes, in every one of which I was entertained with yams, cocoa-nuts, and ſweetmeats. I was afterwards conducted to the King, to whom I ſignified, by ſigns, that I wiſhed much to return to my brother: he underſtood me perfectly, and explained to me, by ſigns alſo, that the canoes could not go out, on account of the great wind. I ſpent the remainder of the day in walking about the iſland, and obſerving its produce, which conſiſted chiefly of yams and cocoanuts: the former they cultivate with great care, in large plantations, which are all in ſwampy watery ground, ſuch

as

as the rice grows in, in India. The cocoa-nuts grow chiefly about their houses, as does also the beetle-nut, which they chew as tobacco."

The favourable account which Mr Wilson brought, joined to the message the King had sent to the Captain by his brother and son, put all our people into great spirits; so that they applied to their several avocations with redoubled vigour, and particularly to getting every thing they could from the wreck.

The number of the visitants increasing very fast, Captain Wilson judged it prudent to set a watch, their guests being previously apprised, lest the turning out suddenly under arms should occasion an alarm. Captain Wilson had kept all his crew under a regular course of exercise, so that they were very expert in handling their musquets, to the no small surprise of Raa Kook and the natives, who were thus impressed with a very favourable opinion of the power and abilities of their visitants.

Arra Kooker, as has been already hinted, was a most extraordinary character; he possessed expressive features, that conveyed his meaning, though totally ignorant of his expressions. He was a man apparently about forty years of age; quite plump or rather round; he attempted at one time to wear trowsers, but found them very disagreeable; but was remarkably fond of a white linen shirt. One of the dogs was his great favourite, and whenever he approached him, the animal leaped and frisked about with every demonstration of joy. His talents for mimickry were so great that he soon discovered many singularities about the strangers, whereby by they were diftinguifhable from one another, and kept them all in fpirits.

Thus, by the kindnefs of Providence were thefe unhappy men, brought from a ftate of the greateft diftrefs, to a fituation at leaft tolerable; indeed, had it not been for the dread of not accomplifhing the building of another veffel to carry them to China, and thus being for ever deprived, in all human probability, of feeing their friends, their families, and country, they might have fpent their time very comfortably.

Chap. II.

The King Vifits them—Reception and Behaviour—Examines every thing, and is pleafed—Farther Proceedings.

ON the 15th, the Englifh were informed that the King was coming; and foon after they faw a great number of canoes

canoes turning the point which formed the harbour; but the King ſtopped as ſoon as he got within the bay, and directed one ſquadron of the canoes, which were all armed, to retire to the back of the iſland; thinking, probably, that ſo great a number of armed people would create an alarm among the ſtrangers. He then came forward with the reſt, in great form, and with much parade, as far as the tide, which was then low, would permit them; and it was ſignified to Captain Wilſon, by the King's brothers, that he ſhould then go and meet him. Accordingly two of his own people took him on their ſhoulders, and carried him through the water to the King's canoe, which he was requeſted to enter; and he and the King, whoſe name was Abba Thulle, embraced one another. The Captain then related the

nature

nature of their misfortune to Abba Thulle, by means of the two Malays, and repeated his requeſt to be permitted to build a veſſel to carry them home; and the King again gave his permiſſion for them to build it, either where they were, or at the iſland where he reſided; but recommended the latter, adding, that the iſland on which they had landed was unhealthy, which was the reaſon it was not inhabited; and that he apprehended they would be ill when another wind began to blow. The Captain informed him, that they had a perſon with them whoſe buſineſs it was to cure diſeaſes; and that it would be very inconvenient to them if they removed farther from the wreck of their veſſel, becauſe they could not then procure from her ſuch things as they might want, without much trouble and loſs of time. To

. theſe

these reasons the King assented; and making signs that he wished to land, the Captain was carried on shore by his people, and Abba Thulle, stepping into the water, followed him. On his landing, he looked about him with a good deal of apparent suspicion, which however was soon removed. Raa Kook made up to him, and a sail being spread for him agreeable to their practice, the chiefs of his company sat also down forming a square; and his other attendants, to the amount of about 300 inclosed them in a circle, squatting down at same time in such a position as that they could rise in a twinkling. Captain Wilson made him a present, of a piece of cloth and some ribbons, which seemed to please him very much. He was quite naked, as well as his brothers, and without any bone on his wrist, or other ornament; he carried a hatchet

of

of Iron on his shoulder, which was so adapted to it, that it gave him no inconvenience.

Abba Thulle, the King, was introduced by Captain Wilson to the officers and all his men, and upon being told that Mr Benger was second in command, he designed him the *Kickaray Rupack*, supposing Captain Wilson to be the King of some country; but when he was made to understand, that he belonged to a mighty Sovereign, and that he was only his *Captain*, he readily got hold of the word *Captain*, by which name he constantly saluted him afterwards, and Mr Benger, *Kickaray Captain*. The King then enquired for Captain Wilson's badge of supremacy, which put him to a stand; luckily Mr Benger slipped his ring into his hand, which being produced, and the manner of wearing it shewn, pleased Abba Thulle

Thulle not a little, as it carried some affinity to their ornament of the bone.

Raa Kook, having as before-mentioned, examined every thing belonging to the English, very minutely, took much pains in pointing them out to the King; they went through the tents, in which every thing surprised them; nor did the difference between the Chinese and English escape their notice. Raa Kook at same time gave his brother to understand, that there were many different nations and classes of mankind on the earth; who were frequently at war with one another, as he often was with his neighbouring Islanders. Abba Thulle appeared to despise the Chinese exceedingly because they had no musquets.

But nothing seemed to strike Abba Thulle with more astonishment than the fire-arms, with which Raa Kook
endeavoured

endeavoured to make him acquainted. He expressed much anxiety to see them used, which Captain Wilson ordered immediately to be done. He desired Mr Benger to cause the sailors go thro' their exercise ranked up on the sea beach, being then low water, while he explained their motions to the King. The men went through various evolutions, with great readiness, marching backwards and forwards, and concluded with three vollies. The astonishment and surprise of the natives on hearing the report of the musquets is not easily conceived; indeed, their hooting and hallooing made a noise little inferior to it. Captain Wilson judged it expedient to be guilty of a little profusion of their powder, on this occasion, in order to impress the minds of the natives, with a more enlarged idea of the power of the English; a design which was

fully

fully anfwered by it. But ftill further to fhew them the effects of their fire-arms, Mr Benger ordered one of the live doves which they had, to be let loofe, at which he fired and immediately brought it down, with a leg and wing broken. This furpaffed every thing in their eftimation; indeed, they now feemed to have loft themfelves in wonder and amazement.

Raa Kook was by this time pretty well acquainted with any articles the Englifh had about them, which he took great pains in pointing out to the King, his brother; thofe that feemed principally to draw their notice, were, a Grinding-Stone, which they turned round with great fatisfaction, obferving the effect of it upon pieces of iron; they alfo examined the tents, and the few culinary articles the Englifh had. But the dogs were the greateft

est fund of entertainment to them; with whose barking they were so much delighted, that they kept a continual uproar with them, and it was found necessary to confine them. The King also examined the English as to their provisions, and got a piece of ham and a live goose.

Abba Thulle was vastly pleased with what he had seen, and proposed going away. This was notified to his attendants by a loud shriek from one of his officers, which gave not a little alarm to the English; it was instantaneously obeyed; the whole rushing to their canoes with great alacrity, and the King with the greater part of the natives took leave of them.

Raa Kook remained with the English all night, as did the King's son and a few of their attendants; Captain Wilson ordered two tents to be pitched, one for the principal people, and the other for the commonalty. He continued himself with Raa Kook and his party, after the guard was set for some hours. The natives in the distant tent anxious to pay all attention to their visitors, prepared to sing a song in their way. Their method of tuning their voices for this purpose, was attended with sounds so very dissonant and harsh, that the English thought they were beginning their war-hoop, or giving a signal to the King and those with him to attack them. Impressed with this idea, every man seized his musquet and ran to the tent where Captain Wilson was, supposing him to be in the most imminent danger. There they were undeceived and attended to the song which was conducted in the following manner; a chief gave out the line, which a company next him took up up and completed the verse; the last line they repeated, and it was taken up by the next party, who also sung a verse. They continued their song some time, and made signs for our people to repay them in kind, which was done by a lad, named Cobbledick, to their great satisfaction. The manner of this lad's singing was afterwards mentioned to the King, who, upon hearing him, was so much pleased, that he never met with him afterwards, without desiring him to sing.

We come now to mention a circumstance, which presents these natives of Pelew in a light that could not have been pre-conceived; a circumstance which discovers such nice feelings, as on the one hand displays human nature in a very pleasing attire, in this her native dress; while on the other, it may put to the blush enlightened nations and individuals. The English had no other means for again revisiting their native homes, but by constructing a small schooner; and for this purpose, had only a few instruments saved from the wreck: these they carefully concealed from the natives, who had shewn a particular attachment to iron and instruments made of that metal. Accidentally, a chief had observed where they were, and requested a cutlass from Captain Wilson: the Captain was loth to part with it, but fearing worse consequences in case he refused, thought it best to give it. As they went out of the tent Raa Kook observed it with great displeasure, took it from him, and returned it to Captain Wilson. In a few hours the *Malay* coming ashore, told the Captain he had given great offence by offering a cutlass to an inferior officer and neglecting the King

King and his brothers. In order to make up matters, Captain Wilſon thought it beſt to preſent each of the King's brothers with ſome cloth and ribbons, which were very coolly received—they ſeemed to be unhappy. In the afternoon, the King came round from the back part of the iſland where he had ſpent the night, and Captain Wilſon went out in the jolly boat with Tom Roſe to meet him. Now Abba Thulle who had been ſo happy and pleaſed the day before, appeared with a gloomy reſerve, and the poor Engliſhmen trembled in anticipating the dreadful effects of his diſpleaſure which they thought he was meditating. But the real cauſe of the uneaſineſs which evidently depreſſed them all, was nothing more than a ſtruggle in their own breaſts, how they ſhould aſk a favour, from ſtrangers almoſt wholly in

in their own power, without having the appearance of a command. A neighbouring nation had injured them, and as they meant to attack them in battle in a few days, they very juftly forefaw the advantages which would arife from the prefence of a few of the Englifh failors with their fire-arms. At length with much evident confufion, Abba Thulle hinted it to Captain Wilfon, who immediately affured him, he might at any time command his men, who were entirely at his fervice. No fooner was this anfwer notified by the interpreter, than every countenance brightened up, and cordiality and happinefs were reftored. The King, immediately dubbed the Captain a brother Rupack, intreated him to fend fome of his people to the part of the ifland where he lived, to carry him whatever provifions they needed, and concluded with
affuring

assuring him, that his people were entirely at his service, to assist in constructing their vessel or any thing else in their power. He immediately retired to the opposite side of the island, promising to return next morning for the men. Nor was the happy settlement of this affair less agreeable to the English than the natives; the fear of having incurred the displeasure of those whose favour was so necessary, had distressed them not a little, and they accordingly were every one more zealous than his neighbour to be chosen for this service; the following five were appointed, Mr Cummin 3d mate, Madan Blanchard, Nicholas Tyacke, Jas. Bluett and Thomas Dutton. They accordingly set off next morning, properly armed and accoutered, Abba Thulle, insisting on leaving four of his attendants by way of hostages.

It

It is a rule among merchant ships, that in case of shipwreck, all superiority of rank or command ceases, and every individual becomes his own master, and may shift as he can for himself; but as they were now preparing to build a vessel, they agreed to form themselves as the people of a dock-yard, and appoint their beloved Captain sole Manager and Director, which was done. Mr Barker who had in his younger years been employed about a dock-yard, designed a plan for the vessel which was agreed to be a schooner, and all heartily and cordially proceeded to the different occupations assigned them; some in felling trees, others carrying them to the dock, while the greater part were engaged in dressing them, &c. Their spirits now revived with the prospect of deliverance, and all seemed but as one man, having only one object in view.

The

The situation of these unhappy people opens a wide field for reflection; many circumstances rendered it very questionable how far it was practicable to complete their schooner; and if they failed, all views of happiness or comfort in this life, they considered as at an end; yet so animated were they with the hope of success, none gave way to desponding, but every one looked forward with rapture to the finishing of their new vessel, as the *summum bonum* which they were capable of enjoying. The Antelope being fixed and stuck to the coral reef, they had opportunity of getting from her various articles which greatly expedited their undertaking. One day they observed a green branch tied to the mast head, this they afterwards learned was a signal to any canoes that might be straggling

gling about, that the King was gone to war, and they ſhould follow.

The boats paid a daily viſit to the wreck, from which they got many valuable articles. They one day brought with them two caſks of beef, and a ſmall caſk of arrack which had been overlooked when the ſpirits were ſtaved. As they now underwent exceſſive labour and fatigue, Captain Wilſon thought it prudent to allow every man a ſmall quantity of grog, which was a valuable cordial to them.

By Sunday the 24th of Auguſt, they had got the keel laid on the blocks, and the ſtem and ſtern poſt bolted. Captain Wilſon therefore ordered every man to attend public prayers and thankſgivings to God for all his kindneſs to them, a practice which was continued every Sunday evening thereafter during their ſtay on the iſland.

Next

Next day they kept a holiday, having already chriftened their embryo of a veffel, *The Relief*, and every man received a double quantity of grog on the occafion.

About this time alfo they raifed a fort of rampart or barricade in front of the tents towards the fea; on the infide they had a foot bank on which they could ftand and fire one of the fix pounders which they had brought from the wreck; and by this means, with the affiftance of two fwivels, alfo mounted on the barricade, they were pretty well prepared to defend the entrance of the cove.

Nine days had now elapfed fince the five Englifhmen had left *Oroolong* with Abba Thulle, and there had been no news from them fince, although it was expected, they would have returned within five days; their friends were therefore

therefore not a little anxious about them. Captain Wilson, uncertain to what cause to attribute their stay, agreed with his comrades in a proposal they made, to settle a plan of defence within the barricade, and to open a communication from one tent to another, through which they could join or retreat in case of an attack.

In the afternoon of Monday the 25th the five men returned, accompanied by Raa Kook, with a present of yams, cocoa-nuts and sweet-meats. Mr Cummin gave the following account of their expedition.

" They spent the night after their departure from the cove, in an island about six leagues to the eastward; and next day arrived at Pelew, which is about four miles distant. Here they spent some days, while the natives were collecting their canoes. On the morning of of the 21ſt they were all aſſembled, to the amount of more than 1000 men, in about 150 canoes; the Engliſh were ſeparated in five different canoes. Early in the afternoon they came in ſight of the enemy, when Raa Kook, having Thomas Dutton in his canoe, went cloſe in by the town and ſpoke to the enemy for ſome time. His harangue they heard with indifference, when he immediately threw a ſpear, which was inſtantly returned; Dutton then fired, and a man fell, to the great confuſion of the enemy. This was ſpeedily followed by a few more ſhots, and Abba Thulle was ſoon left maſter of the field. The flight was all the mark of victory wiſhed for; and all the trophies they deſired, were a few cocoa-nuts and yams. After the engagement, the fleet returned homeward; the King called at ſeveral places by the way, where the women brought out sweet-meat liquor to drink. They did not reach Pelew till seven o'clock next evening, having spent the preceding night in various small creeks on the way. On their landing the English fired a volley and gave three cheers, to the great entertainment of the natives. Public rejoicings took place throughout the island, and every mark of attention and respect was paid to the English. The King took them to his own house, treated them with stewed turtle, thanked them most politely for their assistance, and enquired at Mr Cummin, whether he could have ten men to assist him in another expedition he had in view; Mr Cummin referred him to Captain Wilson as he had no authority, and departed very much pleased with the expedition. They landed on the island where they had spent the first night

night, were again most hospitably entertained, and returned next morning to their companions. This evening they spent with much festivity, the absentees had their nine days complement of arrack measured out to them, and together with the yams and cocoanuts made themselves happier than a few days before they could have supposed it possible for them to be. Mr Cummin described the arms of the natives as consisting of darts from five to eight feet long, pointed and bearded with the wood of the beetle-nut tree, which they use when closely engaged; when they fight at a distance, they use short ones, which by means of an elastic piece of bamboo, they throw at a particular object with surprising dexterity."

On the morning of the 27th, some of the hands were dispatched in the jolly boat to the watering place, and to cast the seine, which they did without success. As the day was pleasant, two or three of them proposed to return over land, and accordingly set out; the remainder returned to the cove in the boat. Night drew on and no tidings of the travellers. People were immediately dispatched with lanthorns to traverse the country in search of them. They hallo'ed as they went along, and luckily discovered the benighted travellers, resting on the very brink of a precipice, over which they must unavoidably have fallen, had not the lights appeared at the critical moment they did.

Raa Kook now informed Captain Wilson, that the King his brother, bestowed on the English the island on which they were, named by the natives *Oroolong;* the Captain accordingly hoisted the British pendant, and took possession seffion for the English; firing three vollies of small arms, as an instrument of possession. Kaa Kook likewise requested Captain Wilson to pay a visit to his brother at Pelew, which from the many things he had to attend to at Oroolong, the Captain was obliged to decline: but he dispatched in his room, Mr Benger the first mate, his own brother Mr Matthias Wilson, and Tom Rose, to compliment the King on his late victory. He also sent one of the Chinese, who are all great botanists, to examine the natural productions of the island. They were received by the King and his people with great hospitality; and entertained with songs in which the word *Englees* was often repeated, seeming to refer to the late engagement, of which they had a grateful recollection; Abba Thulle mentioned to Mr Benger a more formidable expedition

expedition he had in contemplation, in which he expected the affiſtance of the Engliſh. Mr Benger ſaid their houſes were very comfortable, ſurrounded with plantations of yams and cocoa-nuts; they have no corn of any kind, although the ſoil appeared to be very rich. They have no cattle nor qua-drupeds but rats. The Chineſe gave a very poor account of the iſland, in which he found nothing to his mind.

Captain Wilſon now ſet out in the boat, to ſurvey the iſland which he had got poſſeſſion of, the whole circumference of which he judged did not exceed three miles. On the north ſide it is all covered with trees and a ſteep rock hangs prominent upon the ſea. There is a fine ſandy beach on the weſt ſide, as well as a fine plain between the hills and the ſea. The ſouth ſide is rocky like the north; but in ſe-

veral

veral interior parts of the island, especially towards the west, there are evident traces of its having formerly been inhabited.

Chap III.

Captain Wilson visits Pelew—Manners of the natives—Ten Men sent with them to battle—Battle of Artingall—Raa Kook visits Oroolong for more men—Death and funeral of the King's Son.

ON Sunday the 31st of August, Captain Wilson resolved to pay his long intended visit to Pelew; and accordingly went about prayers in the morning, previous to his setting out. Though it did not appear during all the time the English were about these islands, that the natives had any religious ceremonies, it is remarkable that on this as well as several other occasions, they

paid

gave no disturbance to the English when so employed; but paid the greatest attention to what they saw, and behaved with the utmost decency. Mr Devis, Mr Sharp and Harry Wilson accompanied the Captain on this visit; The English in their jolly boat, attended by Raa Kook and other natives in a canoe. As a mark of the uncommon attention which Raa Kook on all occasions paid them, the following circumstance may be noticed. About noon, when they were as yet three or four miles distant from Pelew, he paddled off with all expedition to a little town by the water edge, from whence he brought them, what provisions he could procure, to refresh them. About 1 o'clock they reached Pelew; fired six musquets and fixed their colours in the ground at the end of the causeway where they landed. Raa Kook conducted

ducted them to a house where they waited the arrival of Abba Thulle. Meantime the natives thronged into the house to have a peep at the English, bringing along with them various refreshments and sweet-meats. In a little it was notified that the King was at hand, when, notwithstanding the multitude then present, the greatest silence prevailed. On his arrival, Captain Wilson embraced him as at first meeting, and presented him with a few trinkets, which were very agreeably received.

Abba Thulle now proposed to conduct them to the town, which is about a quarter of a mile from the landing place, where they were. The English, in order to assume some little formality, carried their colours before them. They passed through a wood, and then came to a fine pavement or causeway; there

there are large broad ſtones laid in the middle for the eaſe of walking, and leſſer ones on the ſides; this led them to the town, where they were conducted to a large ſquare pavement, ſurrounded by houſes. In the centre ſtood a larger houſe than the reſt, which was allotted to the Engliſh for their accommodation. In it there were a number of women, of a ſuperior rank, being wives to the Rupacks or principal officers of ſtate, who received them very politely and preſented them with cocoa-nuts and ſweet drink of which all partook.

In a little the King, after a ſuitable apology to Captain Wilſon, retired to bathe, and a meſſage was ſent from the Queen, expreſſing a wiſh to be favoured with the company of the Engliſh at her houſe; thither they all repaired, and were ſeated in a little ſquare before the

the houſe. It appeared that this lady was the principal wife of Abba Thulle, (for he had others,) great attention being paid to her by all; the King reſided almoſt conſtantly at her houſe. She appeared at the window, and by means of Raa Kook, examined into the various peculiarities in the appearance of the Engliſh which ſtruck her. She ſent them a broiled pigeon, which is the greateſt rarity the iſland produces, and is held in the higheſt eſtimation; it is unlawful for any but Rupacks and their wives to taſte them. After ſatisfying her curioſity, they were conducted by the General to his houſe, where they met with a very different reception, and had an opportunity of obſerving the benevolent heart of this worthy man in domeſtic life. In his houſe they were treated with the greateſt kindneſs, and with the moſt expreſſive tokens

tokens of real welcome; but what particularly warmed their hearts on this occasion, was the endearing behaviour of Raa Kook to his wife and children. These last he fondled on his knees and encouraged with all the genuine marks of parental affection. The night was now pretty far advanced, when they retired to their house, where their friend the General spared no pains to render their accommodation comfortable. He procured plenty of mats for them to sleep on, kindled fires to defend them from the mosquitos and damps, and ordered some of his own men to sleep at the other end to protect them from any of the natives, who might be led to disturb them from motives of curiosity. Next morning they were attended as usual by Raa Kook, and after walking about for some time, were ordered to attend the King to breakfast

in

in the Queen's houfe where they had been the day before. They were received with a peculiar etiquette, which was never afterwards practifed. The houfe was all in one apartment; at the one end of which hung a fcreen of mats which when drawn up difcovered the King and Queen feated. They breakfafted on yams and fifh very agreeably. After breakfaft Mr Sharp the furgeon, accompanied by Mr Devis, fet out to vifit a child of Arra Kook's, which was fick. His houfe was about three miles diftant; this gave them an opportunity of examining the country, which they had not before done. This vifit was very acceptable, and the Rupack thought he could not fufficiently repay them. Mr Sharp examined the child's body, which was almoft covered with ulcers, but could not prefcribe any thing, having no medicines. He approved

approved of the mode of cure they had adopted, which was chiefly fomentation. Arra Kook then laded several servants with provisions, &c. in baskets, to be sent to the boats, and assured them when they left the island, they should have his whole rookery of pigeons. This by the way, was the greatest compliment he could offer them, in his estimation, and sufficiently shews the uncommon gratitude with which his bosom was warmed—indeed, the readers will on many occasions have anticipated the remark, that the finer feelings and virtues which adorn humanity, shone in these natives in no common degree. They returned to Captain Wilson at Pelew the same evening.

The request which had been repeatedly mentioned by Abba Thulle, was now formally made to Captain Wilson, by desire of a council of Rupacks, *viz.* that.

that he would allow them ten men to accompany them to a second engagement at Artingall, which was most readily complied with; Captain Wilson mentioned at same time that it would be obliging were the men detained as short time as possible, not to hinder the progress of their schooner; to this Abba Thulle most engagingly replied, " That it was not his wish to detain them longer than was absolutely necessary, but after doing him so much service, he behoved to keep them a day or two to rejoice with him." The council had met in the forenoon on this business; every Rupack or chief was seated on a stone, that for the King being higher than the rest, and disputed from side to side as it happened, without any regular order of speakers; it appeared that every thing was decided by a majority, so that their government bears no small affinity to our own.

The remainder of the time the English spent at Pelew, was very agreeably employed. One day when in company with a great number of the natives, Mr Devis, who was an excellent draughtsman, took out his pencil, and was busily employed in taking the likeness of a woman who drew his attention; the lady observing him, and ignorant of his intention, retired in great confusion. A chief beside him, noticing the drawing, was greatly pleased and shewed it to the King, who immediately ordered two women to come forward and stand in a proper position for Mr Devis to take their likeness. Mr Devis soon finished his sketches and presented them to the King who was highly entertained, and calling the women shewed them their portraits, with which

which they were much pleased. Abba Thulle then desired Mr Devis to give him his pencil and paper, on which he scratched a few figures, very rudely, but sufficiently to shew his conception of what had been done. So that while he thus displayed his own inferiority to the Artist, he at same time gave evident proofs of the sense he had of it, and his wishes to possess these qualifications which so pleased him.

Captain Wilson and his companions were carried to see their method of building canoes, by which means they saw some canoes which were just returned from a skirmish, in which they had proved victorious; they had captured a canoe, which was considered as great a trophy, as a first rate man of war would be in Britain. On this occasion the English had an opportunity ty of obferving their method of celebrating fuch exploits, or keeping a day of feftivity. There was a great feaft prepared for the warriors, previous to which they danced in the following manner. They ornamented themfelves with plantain leaves, nicely paired into ftripes, like our ribbons, which being of a yellowifh colour, had a good effect on their dark fkins; then forming themfelves into circles, one within another, an elderly perfon began a fong, or long fentence, (for they were not certain which,) and on his coming to the end of it, all the dancers joined in concert, dancing along, at fame time; then a new fentence was pronounced and danced to, which continued till every one had fung, and his verfe been danced to. Their manner of dancing, is not fo much capering and leaping, or other feats of agility, as a certain method

method of reclining their bodies and yet preserving their ballance. During the dance sweet drink was handed about, and when it was finished, an elegant supper was brought in.

Mr Sharp carried Captain Wilson one afternoon to see his favourite Arra Kook, who received them with great joy, and entertained them very kindly. They went through many plantations on their way, and were much surprised to find the country so highly cultivated. They observed a tree named by the natives *Ri'a'mall*, which the English supposed to be a species of the bread-fruit. After enjoying plentifully this good man's bounty, they returned to Pelew, highly delighted with their agreeable excursion. In the course of any observations they had opportunity of making, they found the employment of the men generally to be making

making darts, hewing trees, &c. while the women, looked after the yams, wrought the mats and baskets, nursed their children and dressed the victuals.

On Thursday the 4th of September they left Pelew, loaded with presents, and amidst the loud acclammations of a vast number of the natives. They arrived safe at the cove about nine in the evening and found all their companions well and proceeding in their work with the utmost alacrity. The Captain immediately informed them of the request the natives had made for ten men, and every one was anxious to be of the party; at length they were determined upon, and ordered to be in readiness on a call.

Elevated with the prospect of happily attaining that great point to which their most sanguine wishes were directed, there was only one thing which
they

they dreaded, and that was, whether they could find a paſſage with ſufficiency of water to carry them through the reef; this Captain Wilſon ſet out in ſearch of, and luckily diſcovered a narrow opening, where there was about three feet and a half water, ſo that at ſpring tides which riſe about nine feet, they could depend on at leaſt twelve feet, which was conſiderably more than their ſchooner could draw.

In the afternoon of Sunday the 7th of September ſome canoes touched at the cove, bringing with them ſome freſh fiſh, which they bartered for iron; and the following day the King arrived, attended by his brothers, the *Prime Miniſter*, and ſeveral other chiefs; they alſo brought fiſh, eſpecially ſome of a ſpecies, which they had not ſeen before, but which when boiled, proved very palatable. It meaſures about three feet

feet in length, and one foot across, the flesh is very firm like a large cod.

During this visit, the King examined every thing with more attention than before, besides many new objects of surprise were now to be seen. The Smith and his forge proved a matter of great astonishment, never did a conjuror keep an audience in such surprise and consternation, as the smith did the natives with his fire, bellows, and anvil; so enamoured were they with the red hot iron, that they could not be prevented from catching the sparks, though many suffered in the attempt. Nor did the Cooper and his casks escape their enquiries; the dispatch with which he hooped and inclosed a barrel, seemed to them the effect of some supernatural power: in short, every thing seemed to surprise so much, and drew so many spectators, that the workmen
could

could not keep elbow room, and were of courſe much impeded in their work. Captain Wilſon had therefore to uſe many ſtratagems to entice away the chiefs, and Raa Kook, was obliged to interpoſe his authority to keep the natives at a diſtance. The barricade which had been erected ſince the King's laſt viſit, was a matter of much ſurpriſe, they examined the breaſt work with attention, and did not fail to enquire the uſe of the ſix pounder and great gun; the Captain did his beſt to explain the uſe of them, giving him to underſtand, that were the people of Artingall, or any other enemies to approach the cove, they could blow them to pieces; and in like manner by turning the ſwivels, he ſhewed them, that they could defend themſelves by land. This information exceeded every thing they had heard; they talked among themſelves,

and

and by their gestures and attitudes seemed to be lost in surprise. But the great matter which occupied the attention of all on this visit, was the appearance of the new vessel. The King examined every thing about her with the most minute attention, calling his workmen, and desiring them to notice and profit by what they saw. The power and effects of the iron work, and the strong manner in which the whole was bolted and wedged together, surprised the artificers still more than the King; so that poring into every thing with the most inquisitive eye, they were not more amazed at what they saw done, than to conceive how it was possible to complete the work, so as to keep out water and answer the purposes of navigation. They had an opportunity of observing an instance of great superstition on this occasion. In order to get pieces

ces of wood proper for the different ufes for which they were wanted, they had made ufe of feveral different kinds; the natives obferved one kind in particular which they pointed out to the Englifh, and requefted they would not ufe, as it would certainly prove unlucky. Captain Wilfon politely thanked them for their well meant hint, at the fame time affuring them he dreaded no harm. The King and his retinue retired as ufual to the back of the ifland and fpent the night.

Next morning they returned over land, ftill full of the idea of the guns, and entreated the Captain to give them a fwivel along with them on the expedition; this Captain Wilfon fhewed him was impracticable, as they required boats, particularly conftructed for working them. He then begged to fee the fix pounder fired, which was ordered to be done. A scene now enſued which it is not eaſy to deſcribe. If the firing of the muſquets occaſioned the ſurpriſe of which we lately took notice, how much it was increaſed on this occaſion may be conceived. The proceſs of loading was attended to very particularly, but the flame and the following report perfectly ſtunned them. They ſtared at one another for a few ſeconds, then puſhing their fingers into their ears run up and down crying out, the noiſe being much too violent for the drum of their ears, not accuſtomed to ſo loud a noiſe. This however only ſerved to ſtimulate their wiſh for having one of the ſwivels with them, which they thought would ſtrike ſuch Terror into their enemies, as a long courſe of years only would efface.

Abba Thulle in the afternoon repeated his requeſt for the ſwivel, which
Captain

Captain Wilson found great difficulty to convince him it was not in his power to give him; indeed he rather suspected the King and his ministers went away not altogether well pleased with the refusal. Mr Benger therefore who had the command of the party, and had all his men ready drawn up with their arms, ordered them immediately into the canoes, and they set sail. Captain Wilson took every opportunity of getting information concerning the neighbouring islands and their situation; the names of the principal were, Artingall, Pellelew and Emillegue.

On Monday the 15th the party returned from the engagement at Artintingall, all well, though some of them had made a very narrow escape; the canoe in which Mr Matthias Wilson and James Duncan were, had been overset by a sudden squall of wind, whereby

whereby both they and four natives along with them had nearly gone to the bottom. Unluckily neither Wilfon nor Duncan could fwim, but by the vigilance and attention of fome of the natives they were got up into a canoe, having kept faſt hold of a piece of raft nearly two hours. They brought the agreeable news of having effected another complete victory at Artingall, which Mr Matthias Wilfon related in nearly the following manner.

" They reached Pelew the fame night they left Oroolong, from whence Abba Thulle wiſhed they ſhould immediately proceed to Artingall; this however, they aſſured him was impracticable, as it rained hard, and would certainly prove hurtful to the arms; they were all well lodged and entertained. The following evening, they went on board the canoes appointed for them

along

along with the King, Arra Kooker, Ràa Kook and the other Rupacks, and a great number of the natives; the old men, women and children, followed them to the water fide, when they founded conch fhells, to notify their departure to the canoes that were yet in their creeks; thefe foon affembled to the amount of more than two hundred. They proceeded flowly, the greater part of the night, but ftopped at an ifland on the way, and flept on the ground for three hours before day break. They foon reached Artingall, and halted till the fun was fairly rifen, and the enemy had notice of their approach; for, let it not pafs unnoticed, it is an eftablifhed rule in thefe iflands, never to attack an enemy under night or unprepared.

As the King had fome days before fent information to Artingall, of his propofed proposed attack, and at the same time terms of peace, he now ordered a canoe with four men in it to proceed to the island, and enquire whether they were to submit or to fight. Each of the heralds had one of the long tail-feathers of the tropic bird stuck upright in their hair, as a symbol of peace. The messengers soon returned, informing that they refused the terms offered them. Immediately Abba Thulle ordered the conch to be sounded, and waved his chinam stick in the air, the signal for forming the line of battle. Meantime the enemy collected their canoes, but kept close by the shore, shewing an evident disinclination to come to battle. Abba Thulle had dressed himself in the scarlet coat which Captain Wilson had given him, and kept one of the Englishmen in his canoe, the other nine were dispersed through

through the fleet in nine different canoes, armed with musquets, cutlasses, bayonets and pistols.

Finding the enemy would not advance, and their present situation being very unfavourable for the attack, the King ordered a party of canoes to go round a neck of high land, and lie there concealed; he then ordered the remainder to exchange a few darts in their present position, and retreat with apparent precipitancy; by these means he expected to draw the enemy from their shores, and the concealed squadron could then get betwixt them and land, and thus hem them in on all hands. He dispatched his orders with great readiness by means of some very swift sailing canoes, which cut the water with astonishing velocity. His scheme took place as wished. The enemy rushed out to pursue the apparent fugitives, and

and the canoes coming round the high land, furrounded them on all fides. Thofe who fled now turned about, and by means of the few fire arms, threw the enemy into terror and confufion. The noife of the mufquets, their friends dropping they knew not how, and the triumphant haloo which the natives of Pelew fet up, totally difcomfitted them; they retreated with precipitation, rufhed through the canoes that were betwixt them and the land; as there were but few of them, and by that means all efcaped but fix canoes and nine natives who were captured. The victory was however confidered as very complete; it is very feldom that any canoes are taken, and two or three prifoners are generally the greateft number. The very dead bodies are carefully carried off the field of battle, left they

they should fall into the hands of the conquerors to expose them.

The conflict from first to last did not continue three hours; therefore having paraded round the enemy's shores, sounding the conch shell in signal of defiance, and firing when any of them appeared within musquet shot, the King ordered the canoes to be collected and to return to Pelew.

It now becomes necessary to mention a practice totally inconsistent with that humanity, which has uniformly been pointed out as a most remarkable feature in the character of these natives of Pelew. Notwithstanding the entreaties and remonstrances of the English, nothing could prevent the death of the prisoners. The reason assigned for this barbarous practice was, that they had formerly retained them as slaves, in which capacity they soon got acquainted ed with their various stores, the creeks and channels of the island, and somehow or other afterwards escaping, made use of the knowledge of the country they had acquired, in assisting them in their depredations; they had therefore found it necessary to kill every person whom the chance of war brought within their power.

All the prisoners had been wounded in the engagement, and seemed to wait their expected fate with great courage. The principal was a Rupack, known by the bone about his wrist; this they endeavoured to wrench from him, but without effect; he struggled to retain it with singular magnanimity; nor did he quit it but died in the contest. His head was stuck on a bamboo, and fixed before the King's house at Pelew.

Mr Benger took great pains to preserve the life of a poor fellow in the canoe

noe where he was, and kept him fafe for two hours, when one of the King's people, who had been wounded, fnatched the Malay's dagger from him, and ftabbed him, before he could be prevented. Mr Benger obferved that the man died very undauntedly, and feemed while in the agonies of death more impreffed with the appearance and colour of his new enemy, than what he was fuffering. In the boat where Mr Wilfon was, there were two prifoners, one of whom was wounded by a fpear in feveral parts of his body, and the other had his thigh broke. When they go to war they knit their hair in a bunch at the crown of the head, and immediately on being captured, they throw it loofe over their faces, waiting the fatal ftroke. No fooner did thefe two victims fignify that they were ready, than they were ordered to fit down in the bottom of the canoe, which the lame man readily did and was immediately killed; the other refifted for fome time, when one of the natives, fnatching Mr Wilfon's bayonet, plunged it into his body; he lay for fome time weltering in his blood, but never uttered a fingle groan.

They touched at feveral iflands on their way home where the bodies were expofed in triumph; and the inhabitants who were either fubjects or allies, rejoiced with them on the occafion. They brought out fweet drink, and other refrefhments, and feemed to participate in the general triumph. A vaft multitude waited at the landing place of Pelew ready to receive them, loaded with fruit, &c. Great feftivity and rejoicings took place immediately, and the praifes of the Englees refounded in their fongs as formerly."

After

After the return of the party all hands were kept busily employed at the schooner, which was now rapidly advancing. They also continued to send frequently to the wreck, from whence they brought a variety of articles very useful to them. That hunger is an excellent whetter of the appetite, has been often experienced, but never more justly than at present; they discovered about 20 bags of rice in the wreck, which having been so long under water, would not now boil to a grain, but a jelly, yet they considered it as very savoury food.

The men who returned from the last expedition brought a message from the King, informing Captain Wilson that he would pay him a visit in a few days, to make his acknowledgements for the assistance he had given them; he at same time retained Tom Rose to give him

him information as to several particulars concerning the English, with which he wished to be acquainted.

Nothing new occurred for several days; the weather was exceedingly rainy and stormy, accompanied with frequent and loud thunder. They made several attempts to recruit their stock of provisions by fishing, but always in vain; whether it was owing to their ignorance of the proper places, or not using proper bait, Captain Wilson never could determine.

On the 22d of September, Tom Rose returned from Pelew, bringing with him a quantity of yams, a jar of mollosses, and a particular apology from Abba Thulle, for having so long delayed his intended visit; but many of the neighbouring Rupacks having come to Pelew, to congratulate him on his late victories, he could not leave them without

without giving offence; and he could not bring them with him left it should be inconvenient to the English. There was something peculiarly delicate in all Abba Thulle's behaviour; what education, or refinement could have suggested more real politeness than this message conveyed; anxious on the one hand to avoid any appearance of ingratitude; and on the other, fearful left his visit should prove prejudicial to those who had served him.

On the evening of the 28th Raa Kook arrived at Oroolong, accompanied by two chiefs of some neighbouring islands; they brought a present of cocoa-nuts, yams and molosses. Soon after their arrival, Captain Wilson read prayers, as was usual, every Sunday evening, at which Raa Kook and several of the natives attended; some of them began to talk aloud, and were immediately checked

checked by Raa Kook, who behaved with great decency; while thus employed, the Malay arrived from Pelew with a meſſage to the Captain, which however Raa Kook would not ſuffer him to deliver, till prayers were ended. Abba Thulle delighted with the ſucceſs he had already obtained by means of the Engliſh, was eager to take advantage of them, while in his territories, to aſſiſt him in ſubduing his enemies; he therefore deſired Raa Kook to requeſt fifteen men, with one of the ſwivel guns to go with them in a third grand expedition. Captain Wilſon having ſeveral things in his mind which had of late given him ſome uneaſineſs as to the behaviour of ſome of the natives, determined to take this opportunity of ſtating his grievances to Raa Kook, which he accordingly did. He complained of ſeveral thefts which the

the natives had committed from the wreck; particularly paper, copper, and a sixpounder; he mentioned the inhuman practice of killing all the prisoners, which made the English regret that they should have any concern in their engagements; and lastly, he told him that he was informed Abba Thulle expected the same homage from him and his countrymen which was paid him by his own people. This circumstance had been privately suggested to the English, for no other reason, as it afterwards appeared, than to occasion a difference if possible between the English and the natives of Pelew. Captain Wilson likewise hinted that he considered it as a piece of disrespect to send home his men without a Rupack, or some person of consequence to attend them. It is not easy to describe the appearance which Raa Kook's countenance

nance assumed on this information.—Shame, vexation and disappointment were each in their turns depicted on it, in a manner much more expressive than any language he could have used. At length after some considerable pause, he assured the Captain, that he would entirely remove any grounds of distrust betwixt them—That as to the articles taken from the ship, they would all be returned, except the paper which had been rendered useless by the rain; the sixpounder had never been taken with an intention to be kept, but that Abba Thulle had sent for it to be shewn to some of his visitors as a curiosity, and meant certainly to return it. The return of the English without a Rupack had been entirely owing to Mr Benger's hurry, who would not delay his departure a little till things were prepared. He repeated what has been formerly mentioned,

mentioned, as their reason for putting the prisoners to death, being a matter not of choice but necessity; at the same time, he agreed to put the prisoners into Captain Wilson's hand to be treated as he should think fit; but what principally distrest him was the idea that his brother should have been represented as expecting homage from the English; this he reprobated as an infamous falsehood. It afterwards appeared that the Malay had been confined some time for his ingenuity in this lie. This good man's reasoning had a very satisfactory effect, and Captain Wilson having first consulted with his officers, agreed to grant their request, only limiting the number of men from fifteen to ten, as more could not be spared from the work.

Matters being thus agreeably concluded, they sat down to supper with great

great pleasure; after which Raa Kook told Mr Sharp he had now a request to make to him, which he hoped he would grant him; and that was, to go along with him to Pelew, to inspect his son's foot, which was very dangerously hurt by a spear, which having sunk deep into the foot was broke off in attempting to pull it out; and the barb of the spear having got in among the small bones, they could not extract it. Meantime, his foot swelled amazingly, to the great distress of the young man. One of the natives, reputed among them as a man of skill, began to cut away the flesh; but after mangling his foot in a terrible manner, he was obliged to desist, as the effusion of blood became so great that he could not continue the operation. They therefore had recourse to their mode of fomentation, of which Mr Sharp much approved, and desired it

to

to be continued till he saw him, which he could not propose at this time, three of the ablest men being sick.

Next day about noon, Raa Kook set sail with ten men under the command of Mr Cummin, and in their absence the remainder continued their labours at Oroolong with unremitting assiduity.

Although not directly in the course of the narrative, as the reader's curiosity will no doubt be somewhat raised to learn the sequel of the young man's history just mentioned, we shall next introduce Mr Sharp's account of his excursion to see him.

Immediately on his landing he went directly to the General, his father's house, who met him with visible distress in his countenance.

Mr Sharp acquainted him, that he was come to see his son, and had brought such instruments with him,

as

as would enable him, he hoped, to administer relief. He smiled approbation, and conducted him to his house, where Abba Thulle, and several of the principal people were assembled. After paying his respects to them, Mr Sharp was informed, that during Raa Kook's stay at Oroolong, the swelling had subsided by means of the fomentation, and they had forced the spear through his foot, as the only method of extracting it. At this time the whole army was setting out on the grand expedition, which the young man hearing, could not bear the thoughts of being absent from. He therefore insisted upon being carried to his canoe, where though he could not stand on his feet to fight, he could raise himself so much up as to throw a spear. He therefore went along, and very early in the engagement fell a sacrifice to his magnanimity;

ty; a spear entering through his throat, occasioned his immediate death. It is impossible to pass over in silence the unhappy fate of this gallant youth. A spirit more truly heroic, history has not left on record; nor need we hesitate to say, that there was more real valour displayed in this action, which accident only has brought on record, than in many feats which have attracted the admiration of many generations.

This also gives us an opportunity of mentioning their mode of burial, to which Mr Sharp was witness on this occasion. Raa Kook desired Mr Sharp and the boatswain to accompany him to the water side, where two canoes were waiting, into which they went, accompanied by about 20 Rupacks, whom they had not formerly seen, as they belonged to another island, tho' friendly to Abba Thulle. Mr Sharp knew

knew not whither they were going, but suffered himself to be conducted by his friend. They landed upon an island about four miles distant from Pelew. They went a little way up into the island, to a small uninhabited village where there were four or five houses, surrounded by a neat pavement. After resting about an hour here, they set forward to a town about half a mile distant, where a great many people of both sexes were assembled, and an entertainment prepared. Immediately after this, the women retired; and in a little, their attention was drawn to the sound of distress and weeping at a little distance; the voices appeared to be principally those of women; Raa Kook immediately led Mr Sharp from the company to the place whence the noise proceeded. They found a great multitude of women attending a dead corpse, which

which was neatly wrapped in a mat, and supported by four men; they kept up a constant lamentation, and were just about to lay it down, when the strangers joined them. The body was immediately deposited in the grave without any ceremony, while the men who had borne it on their shoulders, proceeded to cover it quickly with the dust. The women then kneeled down, and their cries increased so much, that they appeared as if they were anxious to tear up the very body again, which had been just buried. A heavy shower of rain obliged Mr Sharp to leave this interesting scene, to seek shelter, but he never could learn the cause of Raa Kook's behaviour on this occasion; as notwithstanding the uncommon regard he had for his late son, whose body they were convinced it was, he preserved the most profound silence

silence on the subject; nor did he appear particularly interested. The most probable conjecture they could form was, that he considered it to be below that dignity of mind which he on all occasions wished to support, to appear concerned on an occasion which generally produces those feelings that betray what they consider as human weakness.

The night proved very stormy, so they could not return to Pelew, but spent the evening with Raa Kook. In the morning Raa Kook carried Mr Sharp and the boatswain, to a little hut contiguous to the place where his son had been buried. Here they found only an old woman, to whom the General spoke for some time; she then went out, but returned in a little, bringing with her two old cocoa-nuts, some red ochre, and a bundle of beetle-nut

with

with the leaves. He took the cocoanuts and crossed them with the ochre, placing them one on each side by him; after which he repeated something to himself, which they supposed to be a prayer; he then crossed the beetle-nut in the same manner, and sat musing over it a little, when he gave them to the woman, who carried them out, as Mr Sharp supposed, to the grave; he wished to follow her, but as Raa Kook appeared under great agitation and not inclined to rise, he did not leave him, nor enquire farther.

Mr Sharp entertained his friends with the inspection of his watch and surgical instruments, with which they were greatly pleased, as well as with the description he gave them of the mode of amputation, &c.

Their countrymen they had left at Pelew were in great distress about their absence;

abſence; they had been witneſſes to the funeral of another young man who had been ſlain in the ſame battle. As they were accidentally ſtraggling thro' the fields about two miles from Pelew, they obſerved a great number of the natives going towards a village, with Abba Thulle at their head. They came to a large pavement, where the King was ſeated, and a great crowd ſurrounded him. Thoſe who bare the corpſe, moved ſlowly on before the King, who addreſſed them in a ſpeech, probably recapitulating the qualifications of the deceaſed. This Eulogium he delivered with great ſolemnity, and the reſpectful ſilence of all around him, added a degree of affecting grandeur to the ſcene. The body was then carried to the grave, attended by women only, and thither Mr Matthias Wilſon followed. He obſerved an aged woman

getting

getting out of the new made grave, whom he fuppofed to be the mother or fome near relation of the deceafed, who had been examining if every thing was properly prepared to her mind. The laft offices they always commit to the women, as the men who are nearly interefted or, relations, might be led to difcover fome exterior marks of grief, which they confider as derogatory to the dignity of manhood. Immediately on the body being laid in the grave, the women fet up loud lamentations, as in the cafe of Raa Kook's fon, and Mr Wilfon left them.

Their graves are made in the fame manner as in this country; fome have a flat ftone laid horizontally on the the grave, to prevent any perfon from trampling upon it. They have alfo particular fpots of ground fet apart for the purpofe of burying their dead.

CHAP.

Chap IV.

Difaster of the Veffel—Account of the Third Engagement—Captain Wilfon's Vifit to Rupacks—Abba Thulle Vifits the Englifh with his Wife and Daughter, &c.

THE Veffel was now confiderably advanced, when an unlucky accident had nearly baulked their high raifed expectations. One night the tide rofe to a very uncommon height, and had nearly wafhed away the blocks from under her. At this time they were very fhort of hands; befides thofe at Pelew, three were very fick, fo that the repairing the accident, and raifing a ftrong bank to defend from any fimilar tide, took up feveral days. The weather was at this time very ftormy and difagreeable, in fo much, that fome days they could not leave the tents to work.

The

The jolly boat was difpatched to Pelew for provifions, and in three days returned, with the agreeable news that the Englifh were fafe returned to Pelew from the expedition to Artingall, which had been very fuccefsful; but Abba Thulle would not yet part with them, as he was anxious to fhew them his gratitude, by entertaining them in the beft manner he could. They now alfo brought with them, the Ship's coppers, that fome of the natives had carried off, on their firft vifit to the wreck; this coming to Raa Kook's knowledge he had ordered them to be returned, as he would by no means fuffer any thing to be kept, that belonged to the Englifh.

On Tuefday the 7th of October all the warriors returned, in high fpirits with the entertainment they had met with at Pelew. They were attended

by

by Raa Kook, who brought with him two jars of moloſſes and ſome excellent yams. They gave the following account of the third engagement.

"The canoes were aſſembled and proceeded on the expedition in the ſame manner as formerly, but were much more numerous. On their arrival at Artingall, the ſame notice of the attack was ſent, but no canoes were to be ſeen, nor any appearance of oppoſition. Raa Kook therefore took the command, and having landed the troops, led them up into the country; while Abba Thulle continued in his canoe, and diſpatched his orders to the two commanders Raa Kook and Arra Kooker. They ſoon met the inhabitants, who defended themſelves with the greateſt reſolution and bravery; the King entreated the Engliſh not to land, leſt any of them ſhould meet with harm; but they, obſerving

serving their friends rather hotly handled, jumped on shore, attacked the enemy, and surrounded a house to which severals of them had retreated. The musquets soon put them to flight, and set the house in flames; upon this one of the Pelew people, regardless of the danger, ran in among the flames, and snatching a burning faggot, carried it to another house, where many of the enemy had taken shelter, set it on fire, and returned safe to his companions. Abba Thulle publickly acknowledged his valour, by putting a string of beads in his ear, and afterwards creating him an inferior Rupack. The English were frequently in considerable danger from the spears of the enemy, which they showered upon them in great numbers; but they were soon dispersed by a round of musquets, by which many of them lost their

their lives. Arra Kooker and Thomas Wilson made a very narrow escape; Arra had ascended the hill in pursuit of the enemy by much too far, when noticing one of the Artingall people coming down, he skulked among some bushes, till he was past; then running after him, fetched him such a blow with a wooden sword as immediately stunned him; but as he was dragging him prisoner to a canoe, Wilson luckily observed three or four of the enemy in pursuit of him; who would in all probability have killed him in a few minutes; he therefore immediately levelled his musquet at them, which they perceiving, instantly turned about and fled; this was a very fortunate circumstance, as although Wilson had presented his musquet, he could do no execution with it, his ammunition having been previously expended.

Five

Five canoes were burnt in this engagement, and the caufeway or landing place demolifhed. The only trophy of victory, the conquerors carried with them, was the large ftone on which the king fat in council; a circumftance which naturally reminds a Briton, of the coronation ftone, which Edward I. carried to London from Scone. Great rejoicings took place on their return to Pelew; though the untimely end of Raa Kook's fon, with another blooming youth, who loft his life in the engagement, ferved to throw a damp on many."

About this time Mr Barker had a very fevere fall, which confined him fome days; but notwithftanding of the delays occafioned by ficknefs and accidents, the fchooner advanced pretty quickly, fo that by the middle of October

October her beams were all laid, and many of them secured.

Matters being in an agreeable train at Oroolong, Captain Wilson now resolved to pay a visit with Abba Thulle, to some of the neighbouring islands, his allies; he was attended by his son Henry Wilson, Tom Rose and Thomas Dutton. They left Oroolong in the morning of the 8th of October, along with their good friend Raa Kook; they reached Pelew by ten at night, where they were kindly received by Abba Thulle's eldest son QUI BILL; the King having set off only a little before for an island called E-mungs: Raa Kook proposed to Captain Wilson to follow him immediately, but as he found himself a good deal indisposed, he rather wished to spend the night at Pelew. They embarked next morning having in their party, two wives

wives of Raa Kook's and Qui Bill. They steered about twelve leagues to the northward, and about noon were off the mouth of a rivulet which runs up into the island of Emungs. Raa Kook then sounded conch shells to notify their arrival. This rivulet they found very difficult to navigate, being both shallow and narrow, besides a number of sharp coral stones on the sides; so that the boatmen had frequently to get out and haul them up. They advanced upwards of a mile before they saw any houses or inhabitants, when upon the conch-shell being again sounded, four young men appeared, who soon retired precipitately as if terrified; in a little time however, a great number of the natives came to the water side, when Captain Wilson and Raa Kook landed, and were conducted to a large house, where

great

great multitudes furrounded them, gazing with much aftonifhment, on the ftrangers, whofe colour and drefs were fo new to them; befides, their curiofity had been raifed by the accounts of their valiant deeds in battle, of which they had heard. They remained here about half an hour, and then fet forward to a large houfe or public building, about a quarter of a mile diftant, where Abba Thulle, and a number of Rupacks were waiting their arrival. After fpending about two hours there, they went to vifit the Rupack of the town, an old infirm man; here they remained about half an hour, and had fome boiled yams, fifh, and fweet drink fet before them; they ate a little and returned to the great houfe, where a great entertainment was waiting them. The company were divided into two parties; in the one were Abba Thulle, Captain

Captain Wilſon and his attendants, and in the other **Raa Kook, Qui Bill** and others. Captain Wilſon here remarked, that nobody ventured to taſte the meat, till the King had given the word, and in like manner none ventured to lie down for ſleep, till he was covered with his mat.

After eating, the natives began a dance in their uſual manner, which continued the whole night; they ſung alſo a good deal, and as both men and women joined, they produced together a terrible noiſe. The Engliſh ſpent but a very uncomfortable night here; they had only rough uneven boards to lie on; and had they been more agreeably laid, the noiſe of the ſinging made it impoſſible to get any reſt. Theſe amuſements continued part of next day; and in the afternoon, they were entertained with a mock fight betwixt two of the natives, and a dance with spears, in their hands, which continued about an hour; during this dance, the chief presented Captain Wilson with four different spears, and a curious wooden sword, inlaid with shell. Next day they had new dances, which however, were soon stopped by a terrible storm of thunder and lightning. The weather clearing up in the afternoon, the old Rupack was brought out to the large pavement, carried on a board flung with ropes on two poles, which were supported by four men. A piece of etiquette now took place, which the English did not understand; all the Rupacks seated themselves with much respect on the pavement, where the old Rupack sat, but Abba Thulle went to a little distance, and sat at the foot of a tree making the handle of a hatchet; his place was supplied among the Rupacks, by Raa

Raa Kook who perfonated and fpoke for him. Having converfed together for fome time, the old Rupack diftributed prefents of beads among the reft. Thefe beads are in general a kind of coloured earth, which they bake by a particular procefs which the Englifh could not get an opportunity of feeing; fome of the Pelew people however had made a parcel out of fome bottles they had got from the Antelope, having turned them very neatly. The method of diftributing thefe beads was as follows: the old Rupack gave fome of them to one of his attendants, who went into the middle of the fquare, then mentioned the perfon for whom they were defigned, pronounced an eulogium upon him, and running up to him, delivered them to him. After the Rupacks had got their fhare, Tom Rofe was fent for Captain Wilfon, who was in the houfe observing

observing the ceremony from a window; to him the old Rupack gave a string of red beads, and two tortoise-shell spoons. After this ceremony, they spent another hour in conversation together, when the old Rupack was carried back again on his board, and refreshments set before the Rupacks.

Captain Wilson took an opportunity of enquiring the reason of a number of human skulls being placed upon the outside of the doors, windows and ends of the great house, and was informed, that not many months before, while the principal people of Emungs were absent on a visit to a neighbouring island, the inhabitants of Artingall landed and attacked the town, putting to death such as could not make their escape; setting the houses on fire and destroying wherever they came; notice of this having reached Pelew, Abba Thulle

Thulle quickly assembled his canoes, and beset them unexpectedly; at the same time, the people of Emungs returning from their visit, they so completely surrounded the Artingall people, that very few of them escaped; and those were the heads of some of the chiefs.

Captain Wilson having signified a wish to depart, next morning the conch-shell was sounded, and the canoes assembled by day-break. About eight o'clock they embarked in company with one *Maath*, a Rupack of consequence in a northern island, who had eight or nine canoes in his retinue; they parted with him at the mouth of the rivulet. He carried with him the scarlet coat and spaniel dog which Captain Wilson had given Abba Thulle, in order to shew them to his countrymen, but they were afterwards returned to Pelew.

Pelew. Before his departure, he anxiously entreated Captain Wilson and his company to go with him, which they declined, as it would have detained them too long from Oroolong. They continued their route homewards very agreeably, till about ten o'clock, when a very violent storm of thunder and lightning, accompanied with a deluge of rain overtook them. The high wind soon dispersed the canoes, and the boatmen of that in which Captain Wilson was, having in vain endeavoured to steer on their way, agreed to make for the shore, which was done; and then kindled a fire by rubbing two sticks together. Raa Kook covered himself with his mat, and his two wives sheltered themselves under his boat cloak, at every flash of lightning, ejaculating what the English supposed to be a prayer. Captain Wilson observed on this and

and other occasions, that the natives of these islands had a peculiar dislike to their skins being wetted by rain, probably the spattering of a shower was disagreeable, for it could not proceed from any aversion to water, as they bathed every day. The weather cleared up about noon, when the companies of the different canoes met together and dined. They then walked up the country to a town named Aramalorgoo, where they were kindly entertained. Before they embarked, Raa Kook expressed a desire to fire a musquet, which he had never done, and Captain Wilson humoured him; but holding it loosely, it struck his shoulder so forcibly, that it fell from his hands and he tumbled backwards. It astonished him to see the English fire it so easily, while he could neither hold it nor stand when fired.

They

They touched at a place called Emelligree, which appeared to be a diſtinct government. The Rupack, a luſty, good-looking old man, ſent them a formal invitation to viſit him, which they did. Abba Thulle deſired that all his company ſhould attend this Rupack, but he himſelf kept by his canoe, probably from the ſame etiquette which regulated his conduct at Emungs. At this place they were treated with all hoſpitality, not only in public, but at ſeveral private houſes; and were detained ſo long that the inhabitants had to conduct them to their canoes with torches, for it was very dark.

It was ten o'clock before the canoe in which Captain Wilſon and his party were, arrived at Pelew; and as the King was not yet come, none of the natives would land; the Engliſh tho' under no reſtraint, choſe to ſhew their politeneſs

politeness on this occasion, by waiting for him also. He soon arrived and spent this night with the English in the house by the water side. Next morning at breakfast, Abba Thulle informed Captain Wilson that he was directed by a Council, to request the assistance of his men in battle, yet once more. The Captain replied, that nothing could give him greater satisfaction than to be of any service to the Pelew people that lay in his power, but when he left Oroolong, so many of the men lay sick, that till he saw how they were, and conversed with his officers, he could not give a positive answer; with this they appeared very well satisfied, and about eleven o'clock in the forenoon, Captain Wilson and his party, with Raa Kook, set sail for Oroolong, where they arrived safe, about four o'clock.

Among the firft employments after Captain Wilfon's return was Difcipline. During his abfence, the Cook had mifbehaved exceedingly, appropriating great part of the fmall portion of the meat they were allowed, to himfelf and his affiftant. As it was neceffary, in their prefent fituation, that the ftricteft difcipline fhould be exercifed, Captain Wilfon, by a Court Martial, ordered him a *cobbing*. The native tendernefs of Raa Kook's difpofition appeared eminently on this occafion; when he faw the man ftripped to the waift and his hands tied againft a tree to keep him extended, he entreated Captain Wilfon to let him off. The punifhment of cobbing is inflicted by a thin flat piece of wood like a battledore; which Raa no fooner faw exercifed, and the man bearing it patiently, than he was reconciled ftanding by by and encouraging him all the time. A Chinese was also punished in the same manner, for wounding one of his countrymen with a stone; but he roared and bellowed so lustily, that Raa Kook was greatly entertained with his cowardice.

The canoes now frequently touched at Oroolong with fish, so that the English had in general plenty of fresh provisions. On the 15th of October, three canoes landed, in one of which was a woman, the first they had yet seen at the cove. She went through the different works and surveyed every thing with great attention, but with great caution. The men that accompanied her did not land, nor could Captain Wilson conceive from whence they came; it was conjectured they were from Emilligree, as none of the English could recollect any of their countenances.

Abba Thulle arrived about ten o'clock on the 17th, with the agreeable news, that the Chief Minifter of Artingall had been at Pelew with offers of peace, which had been concluded upon, to the great joy of Raa Kook and the other natives. Abba Thulle brought his youngeft daughter with him, named *Erre Befs*, of whom he appeared to be exceedingly fond; he conducted her through all the cove and explained the ufe of every thing with much attention. Befides her he alfo brought with him on this vifit *Ludee*, one of his wives; a very beautiful woman, young, and greatly fuperior to any of the females they had hitherto feen; her genteel deportment and graceful ftep drew the attention of every beholder. She had with her eight or ten females, who were all efcorted by Raa Kook, and fhewn the forge, veffel, guns, tents and other

other curiosities, with which they were greatly surprised. The King had also brought some of his artificers with him (or *Tacklebys*, as he called them) to observe the progress of the vessel, &c. he seemed peculiarly anxious that they should pay attention to the schooner, which all ranks agreed in considering as the *ne plus ultra* of human workmanship. After their curiosity had been fully satisfied, the Captain prepared an entertainment for them in the tent, consisting principally of fish, and boiled rice, sweetened with molosses, of which they appeared very fond.

A good deal of conversation took place on this visit between the King and Captain Wilson on various subjects. Abba Thulle acknowledged that the English musquets had now procured him peace with almost all his neighbours; he at same time requested that the Captain would leave ten musquets with him when he left the island; this Captain Wilson told him would not be in his power, as Britain was at present engaged in war with several different nations, with whose vessels they might fall in on their return homeward, and so require defensive weapons; but he promised him five, which greatly pleased him. Abba Thulle then enquired what quantity of powder they had, but observing that Captain Wilson was not disposed to answer him readily, he very politely changed the subject. The Captain then desired he would assure his neighbouring islanders, that the English, deeply sensible of the kind usage they had received from the inhabitants of Pelew, were determined to return very soon, in a much larger ship, and with a greater number of men, and fully avenge any insult that

that might be offered to the Pelewites, either by the people of Artingall, or any other island.

Agreeable to a former promise of Captain Wilson's, Abba Thulle then informed him, he had come at this time to get the guns from the wreck, which should either be placed at Oroolong or Pelew, as the English pleased; Captain Wilson having previously consulted his officers, desired him to take them all to Pelew, except one, which they might perhaps need in the schooner. Accordingly, next day, the King ordered some of his people to go to the wreck in order to remove them. Having no tackle, they found it a very difficult jobb, and were forced to send for ten of our people to assist them; the Englishmen speedily lodged them in the canoes to the surprise of the natives, who could not conceive it possible to handle these

these heavy pieces with such apparent ease.

The King lodged at the back of the island, carrying with him all his attendants, that the English might be as little interrupted by them as possible. He had not been long there, when he sent for Captain Wilson to give him ten large fish, part of a quantity his people had taken; of these he would only receive four, which would fully supper all his people, and such is the nature of the climate there, that no fish will keep fresh above five or six hours. The King then ordered the remaining six to be dressed for keeping, and sent to the cove in the morning. Their method of cleaning and dressing them is as follows; the fish is first well cleaned, washed, and all the scales taken off; then two sticks are placed lengthways of the fish, in order to keep it straight,

in

in the fame manner as fticks are placed acrofs falmon in this country when kippering; it is then bound round with broad plantain leaves, and fmoked over a flow fire. In this ftate it will be eatable for at leaft two days, though not very pleafant.

In the morning, the ears of the Englifh were faluted with the noife of finging in the woods, which proved to be Raa Kook and his attendants coming acrofs the country, with the fix dried fifh, which were very acceptable. This morning the King went to the wreck, and returned to the cove, and breakfafted on tea with Captain Wilfon, three Artingall people being alfo of the party. After breakfaft, the ftrangers were led through the works, and their furprife was nothing inferior to any that had yet been expreffed; the guns particularly interefted them, as the

the means by which so many of their countrymen died, in a manner then incomprehensible. In a few days they had a farther opportunity of seeing the effects of the musquets, by Mr Benger's killing some pigeons while on wing, they run to the carcases, and examined them very attentively, and upon noticing the wounds, observed, it was with such holes as these their countrymen died; on this occasion the Pelewites seemed to exult a little over their neighbours, on the ignorance which they shewed of the use of fire arms. The People of Artingall however retained no animosity on this account but seemed quite happy and at ease.

Captain Wilson had now occasion to complain to Abba Thulle of a theft which had been committed by some of his people, a cooper's adze and a caulking iron being amissing; the latter he recovered

recovered immediately, but the adze he found had been carried to Pelew.

On Monday the 20th the King went again to the wreck, and in his abfence a meffage was fent by Raa Kook to Captain Wilfon, informing him that he was very bad, and wifhed much to fee the Captain and Mr Sharp. They went immediately to fee their good friend, whom they found much diftreffed with a large boil on his arm, which was attended with a confiderable degree of fever. Mr Sharp dreffed it with care, and found him much better on his return in the evening, when the King was prefent, and expreffed great thankfullnefs for the attention paid to his brother; Abba Thulle appeared very much interefted for his recovery, and on every occafion manifefted the highest concern about all his relations and friends. Captain Wilfon obferved

ved when there in the forenoon, that Raa Kook was attended by his wives, who appeared greatly agitated; their breafts were fcratched and bleeding, by means of a prickly leaf, which they applied very fmartly, in order to teftify their concern.

During Abba Thulle's refidence on Oroolong, the Englifh were regularly fupplied with frefh and dried fifh, which were very acceptable; among the reft were fome Kima Cockles, fo famous throughout Europe for their beautiful fhells. Another unknown animal was obferved about this time, fomewhat fimilar to our batt, but four times its fize; it runs along the ground, climbs trees and leaps from branch to branch with great alertnefs; befides which, it has wide extended wings, and flies rapidly. It is efteemed a nice difh at Pelew, and like the pigeon,

pigeon, sacred for chiefs only. On the 21st Abba Thulle came round to the cove on his way to Pelew. He asked Captain Wilson if the English would assist him in battle once more, to which Captain Wilson readily agreed. He then told him that there would be a grand council held at Pelew on the subject the next day, the result of which he would communicate; on this account Raa Kook would not remain behind, though far from well, his presence being necessary in council. The King then informed the Captain, that he would send him a quantity of paint for their vessel, and desired the jolly boat might be sent for it; this was done, and a quantity sent, much more than sufficient for their purpose. It consisted of red and yellow ochre, being all the natural paints of the country; the King sent a strict charge to preserve

serve the baskets, in which the colours were packed, from wet; and informed, that men would be sent proper for painting the vessel, on their return from the proposed expedition. Mr Devis, Tom Rose, and another of the English people went with the King to Pelew, and remained there till the 24th, in which time Abba Thulle had made still farther enquiries at them as to some particulars concerning which he wished to be informed, particularly as to the nations with whom Captain Wilson had mentioned the English were at war.

By the 26th of October the vessel was breamed and the outside caulking completed. The same day, a number of strangers touched at the cove, in ten canoes; they proved to be friends to Abba Thulle, on their way to join his fleet. It appeared they had previously oufly heard of the Englifh, and by that means were not fo much furprifed on feeing them; yet they were greatly entertained with the various works, through which they were conducted. Their chief was an elderly man, and fpent the greater part of the time they were on fhore in Captain Wilfon's tent; finding a book on his table, he was much pleafed with the appearance of it, and entertained himfelf by reckoning the number of leaves in it, a tafk which he could not accomplifh, having frequently proceeded the length of fifty, but could not go further.

Mr Sharp now paid a vifit to Raa Kook, to enquire after his health, and was happy to find him much better. One of the furgeons at Pelew had cut the core of the boil, and the furrounding flefh with one of their knives. When we confider the nature of the inftrument ſtrument with which all their ſurgical operations are performed, and that all their knowledge conſiſts in cutting out the part affected, one cannot avoid feeling for the pain, or rather torture, which thoſe muſt endure, who are ſo unhappy as to need their aſſiſtance. Perhaps they may now be enabled to execute their operations with ſomewhat leſs pain, as the Engliſh diſtributed a few twopenny knives among them. Mr Sharp re-dreſſed Raa Kook's ſore, and left ſeveral dreſſings, with proper directions for application. His viſit to the General was conſidered as very flattering, and warmly recommended him not to the friends of his patient only, but to all the iſland, who were particularly fond of Raa Kook.

In the afternoon of Monday, the 27th of October, Abba Thulle, arrived at Oroolong, with a great number of of canoes in his train; they formed the third grand divifion of the fleet, which altogether confifted of upwards of 300 canoes. They failed in good order and made a very formidable appearance. The following ten men prepared to go on this expedition; Mr Matthias Wilfon, Thomas Wilfon, William Roberts, Thomas Dutton, Nicholas Tyacke, Madan Blanchard, Thomas Whitfield, John Duncan, Jas. Swift and William Steward. Immediately on the King landing he was informed the men were ready; upon which they all embarked, the Englifh on fhore giving them three cheers, which they in concert with the natives returned very warmly.

Chap. V.

Expedition to Pelelew—Visit from Abba Thulle—Launching of the Schooner—Presents to the King—Preparations for Departure—Captain Wilson made a Rupack—Blanchard's Determination to Remain—Proposal for Lee Boo returning with the English—other Proceedings previous to Sailing.—

THE Armament were scarcely out of sight, when a very furious storm arose at Oroolong, which distressed them all exceedingly; not only on account of the danger their vessel and tents were in, but the apprehensions they reasonably entertained for the safety of their absent friends. They were however soon relieved; on Thursday the 30th October about midnight, their companions having been only gone about three days, a canoe was observed coming towards the harbour, and the crew calling out *Englees*, they were permitted

mitted to land. In this canoe, came Arra Kook, and the furgeon's *fucalic,* or friend, the news of whofe arrival foon raifed the Englifh from their beds to hear fome account of their friends. The original caufe of the conteft with the natives of Pelelew it feems had been, their refufing to deliver up two people that had been faved from the Malay wreck; immediately therefore, on Abba Thulle's appearing on their coaft, the enemy laid down their weapons, offered prefents, and delivered up the two *Malay-men.* Next day, thefe agreeable tidings were confirmed by the return of the party, who gave the following account of the expedition.

" They met with very bad weather the firft night, and were obliged to land on an uninhabited ifland, about four leagues diftant from Pelelew, where they erected temporary huts. Next day two
different

different parties went on an excursion to a neighbouring island, where they terrified the natives, and pillaged a little. Abba Thulle then called a council, in consequence of which Arra Kooker set off next morning to Pelelew, and concluded the peace. On his return in the afternoon, the English were informed, that in the present situation of affairs it would be derogatory to his dignity to approach Pelelew, but if they had any wish to see the island, Arra Kooker would attend them. This offer was accepted, having previously entered into a bargain, to keep close together, in case of treachery on the part of the Pelelewans. They were attended by a great number of the Pelew people, and highly entertained with the island. They found the country pleasant and fertile; the land appeared nearly level, and the houses were large

and

and better constructed than those at Pelew: The natives seemed to be friendly and humane, and they shewed a very marked respect to our countrymen, although the object of their visit had been to spread devastation. Indeed the natives of Pelew discovered a degree of rancour against these Pelelewans rather inconsistent either with their character or practice; the English were ready to think they had been stimulated to it by the Malay, in order to get his two friends out among them.

Abba Thulle made a very short stay at Oroolong, as did even Raa Kook; they returned to Pelew accompanied by the King of Pelelew. This Rupack was an elderly man, and of a stern rough appearance; his hair was grey, and his beard tapered to a narrow point; The only other singularity about him, from the people of Pelew was, his being

ing tatooed quite up to the navel, while their tatooing did not exceed the middle of their thighs. Mr Sharp with four of the men fet off the day following for Pelew, in order to bring over fome moloffes promifed them by Raa Kook. He mentioned that the ufual rejoicings had taken place at Pelew on the happy termination of this expedition; and fongs were compofed on the occafion in which thefe words could be diftinguifhed, " *Englees*,—Weel a *Trecoy*" (very good); and in a particular manner *Tom Rofe* was celebrated in them. This agrecable jocofe young man, acted as an interpreter on all occafions, and had made himfelf a favourite among the natives.

About this time Captain Wilfon intimated a wifh to his officers and crew, that they would agree to a few days being fpent in navigating round the fhores shores of this cluster of islands, where they had spent so many weeks with a degree of comfort and satisfaction, far surpassing their most sanguine expectations. He told them they were the first Europeans who had hitherto visited them, and as human nature was here to be seen in an attire, she had never assumed, so far as he had heard, it would be acceptable to their Employers, and a service to mankind, to spend some little time in visiting the other islands at which they had not yet touched; he said that Abba Thulle would chearfully give them all the assistance in his power, and it was a task which could easily be overtaken. But however plausible the scheme, it was approved by none—the fear of being engaged in hostilities with some of those islands to which they were strangers; the danger of encountering sea storms in small canoes;

noes; and above all, the protracting the anxiously expected hour when they should again set sail for their native country, determined them all to entreat the Captain to lay aside the scheme; which he accordingly did.

Meanwhile the vessel advanced apace, and a consultation was held, to fix on the safest method of launching her, which was agreed to be lay ways. They had neither pitch nor rosin to pay her with; this want, necessity, the mother of invention, taught them to supply by burning coral stone into a lime; then sifting it thoroughly, they mixed it up with grease, and found in it an excellent succedaneum.

The Reader will not have followed his countrymen thus far, and seen Providence rescuing them from the jaws of death, and fostering them in these to us unknown regions, without feeling his heart warmed in no common degree to the friendly islanders, by whom they were protected; yet it is now necessary to unfold a scene on the part of the English, which without well weighing the accompanying circumstances, he will be ready to censure as ungrateful, ungenerous, and cruel. Instances of such noble sentiments of liberality, uncontrouled by art or interest, and continued so long, are so very rare even in civilized nations, that the minds of several of the English were ready to take alarm at the smallest accident, lest all this overflow of kindness, should only be to lull them in security, till they should in the first place serve their own ends by them, and then more effectually cut them off. A message from the King at this time,

informing

informing that he proposed paying them a visit in a few days, and remaining with them till their departure, gave the first alarm; but on Tuesday November the 4th, two canoes being observed off the harbour, towards night, and neither of them coming in, it was instantly concluded that they were spies, and that the natives, conscious of the value of their aid in battle, intended to prevent their departure. Every precaution was therefore made for a vigorous defence, in spite of the eloquence of Captain Wilson, who insisted that it was a degree of injustice to entertain even a suspicion of a people so hospitable, and a prince so generous and condescending. He ridiculed the idea of defence, supposing the natives really meant to detain them, as the ammunition was nearly expended, and the natives could so easily cut off

off their frefh water. Thefe and many other arguments equally forcible availed nothing; it was determined that every one fhould be on his guard; that the fwivels and fix pounder fhould be loaded with grape fhot, the fmall arms charged with ball, and the cartouch boxes loaded with cartridges, and left they might be overpowered with numbers, it was agreed to fingle out the chiefs for the bayonet or mufquet, with a view to difmay and difperfe the multitude. There is not one circumftance in all this narrative, which is more difagreeable to relate, than what common fidelity requires to be now mentioned; that the amiable and princely Abba Thulle, the humane, benevolent Raa Kook, and the jocofe, entertaining and warm hearted Arra Kooker, were to have been the firft victims of this phrenzy; not that the

English were deadened to every sense of honour, gratitude, and generosity; but when the precious cup of liberty was now almost at their lips, the dread of its being yet wrenched from them, however groundless the idea, wrought so powerfully as to absorbe, *for a moment*, every principle of honour.—— Happy are we to say, it was but for a little time, that these ungenerous sentiments prevailed: the reflection of a night greatly moderated them, and by morning light, there were none of the company who did not feel, in some degree, the force of what Captain Wilson had alledged the preceding night; yea many among them blushed at the appearance they had made to one another, discovering the weakness of human nature, when struggling between the hopes of returning to their native country

country, and the dread of perpetual detention.

Agreeable to a previous promife he had made to the King, Captain Wilfon difpatched the jolly boat to Pelew, on the morning of the 6th of November, under the direction of Mr Sharp and Mr Matthias Wilfon, who carried along with them, all the iron and tools they could fpare. They were defired to inform Abba Thulle, that until the veffel was completely finifhed, they could not fpare him the mufquets, nor any more tools, but in the mean time, they hoped to have the pleafure of a vifit from him, as they expected to be ready to fail in a few days; they were alfo defired to exprefs in the warmeft manner, the high fenfe the Englifh entertained of the unbounded kindnefs they had received, which they were determined publicly to declare on their return

return to Britain. While Captain Wilson was talking with his ambassadors about this message, Madan Blanchard entered the apartment in quest of some tools he wanted; and immediately took the opportunity of desiring Tom Rose, who was to be of the party, to inform the King that he was determined to stay behind and reside at Pelew; upon the Captain ridiculing his message, he solemnly declared that he would not embark with them. Many attempts were made to convince him of the impropriety of such a step, but all in vain; mean time, the Captain ordered that this circumstance should not be mentioned at present, and the boat departed. By Captain Wilson's desire, Blanchard's companions used every argument in their power to divert him from a scheme so very imprudent in every point of view; but he informed them

his

his mind was resolved. The idea of deserting his comrades suggested itself on his return from the first expedition against Artingall, and he then mentioned it; at the same time adding, that he would cheerfully join in their daily labours, with the same diligence and perseverance as any of them; but that he had resolved to end his days at Pelew, without again encountering the conflicting elements at sea. Having formed no particular connection with any females, they looked upon him as in jest, but he never altered his resolution.

Next day the boat returned from Pelew, having in company, the King, his young favourite daughter, Raa Kook and several chiefs of distinction; they had been impeded by a storm; and as was formerly noticed, though these people go perfectly naked, and frequently

quently bathe, they fly to shelter from a storm of rain, with the eagerness of an English Beau, to preserve a new or fashionable coat. The Englishmen in the pinnace fell in with the Pelew company in their canoes during the storm, and accompanied them to the island of Pethoull, where they spent the night together. The ladies who were of the party, expressed not a little disappointment in being obliged to halt a night by the way, as they were very impatient to see the launch at Oroolong. They supped together very cheerfully, when there was again opportunity of remarking, that no one presumed to eat till the monarch had pronounced the word *Munga*, that is *Eat*, upon which a signal is given to the attendants without, when all begin to eat together. Each one's share was portioned out on a plantain leaf, which
served

served for a plate; though on great occasions they use a sort of dish made of tortoise-shell, and others of earthenware and wood; they cut their meat with a knife made of split bamboo, with which they carve very decently.

Mess. Sharp and Wilson now presented the King with the iron tools, which were very graciously received, they explained the method of using them, to which Raa Kook paid particular attention; the Malay took this opportunity of hinting, that the English had not sent the musquets they had promised; to this Raa Kook with great indignation in his countenance, replied, that they had sent all they had promised to send, previous to their departure, of which they had given due notice; that this insinuation, was like the former, whereby he had nearly effected a difference between them: the
discredit

discredit which this behaviour brought upon the Malay, not with Raa Kook only, but all the chiefs, quite disconcerted him, and he retired in confusion.

The order in which the company sat during this night's entertainment, is worthy notice. The house in which they were was in one large apartment, torches were lighted, and stuck in betwixt the boards of the floor, in a line through the centre of the room; and the company sat in rows with their backs to the light.

When ready to retire for rest, the servants brought in the mats, and fires being lighted to defend from the musquitoes, the torches were extinguished, and all was silence. This however was interrupted about two or three in the morning by the arrival of a messenger, who was introduced to the King. This they

they afterwards learned was a meſſage from ſome of the neighbouring iſlands enquiring the time of the departure of the Engliſh. The method which Abba Thulle uſed to ſend his anſwer, is very noticeable, he took a ſtring, and put as many knots upon it as there were days to the time. This enquiry at firſt ſounded rather diſagreeable to the Engliſh, but it afterwards proved to be a freſh inſtance of their kindneſs, as they only wiſhed to know, for the purpoſe of contributing to their ſea ſtore.

Next morning proved calm, and they ſet ſail; the King and daughter, Raa Kook and another chief, went with the Engliſh in the pinnace; a ſudden ſquall ariſing nearly overſet moſt of the canoes, but the pinnace, ſailed very eaſily to the great ſatisfaction of Abba Thulle, who was mightily pleaſed

to

to hear that Captain Wilson proposed leaving her with him.

Immediately on his arrival, Abba Thulle ordered his men to proceed in painting the vessel, which they did. Raa Kook assisted himself in the operation, under the immediate direction of the King. The stern was the place which he decorated with all his art; in particular he was at pains with two circles, some ornaments, hanging from them, the particular intention of which the English never discovered. In the afternoon, the ship was lowered off the blocks upon the ways; but being too much over to one side, was swept with a rope, and to bouse her over, a tackle got upon it. The King attentive to their proceedings fetched a long pole, to apply as a lever to heave the vessel over; but, on a hint being given that it was wrong, he desisted; she was easily
got

got upon the ways, and every thing being ready, the next morning was appointed for the important launch. The King now sat down near the scene of bustle; and after a short conversation with his brother and the chiefs, ordered it to be signified to Captain Wilson, that it was his request he would change the name of the ship to that of the Oroolong, in remembrance of the island, and the people. A ready concurrence being universally given, the good prince appeared more than commonly happy. The Captain being told that Blanchard was coming to offer himself to the Pelew King, determined to make a merit of necessity, and therefore signified to Abba Thulle, that as a return for the hospitality with which the English had been treated, they would leave one of their comrades with him, as a perpetual residenter, who was qualified to manage

nage the great guns and other things, beyond their comprehenfion. The idea was by no means thrown away; the King was gratified beyond meafure.

This night Blanchard fpent with the King, and was well entertained. He promifed to make him a Rupack, to give him a houfe and plantations, and allow him two wives. All the crew regretted much to part with Blanchard; his agreeable behaviour made him beloved by his companions, who loft no opportunity to fpeak in his behalf to the natives. This refolution of his, however, was inexplicable; as it is difficult to conjecture what motives could urge him to forfake that clafs of mankind among whom he had hitherto lived, and be feparated from them perhaps for ever. As Abba Thulle, Raa Kook and the natives in general, confidered his remaining among them as a very great compliment,

compliment, they were refolved to make him happy; and there is great probability, he now lives among them in a fituation not only comfortable but refpectable. Meantime the reader probably looks forward with anxiety to the return of thefe fhips, which it is expected have ere this, paid a vifit to thefe iflands, in hopes of hearing somewhat concerning the future fortunes of this young man. He was only about twenty years of age when left there.

Abba Thulle defired Captain Wilfon and his officers to point out to him a proper fpot about the cove, where he meant to plant fome cocoa-nuts and yams, for the refrefhment of the Englifh on their next vifit. This was done accordingly, and it was obferved, that upon the grain being put in the ground, the perfon planting muttered a few words to himfelf. An attempt was alfo made to launch the veſſel this afternoon but in vain.

We come now to contemplate a ſcene peculiarly intereſting. Next morning (Sunday the 9th of October) the Engliſh proceeded before day-break to make ready for the launch; it need ſcarcely be mentioned, that uncommon pains were taken to put every thing in the moſt favourable train for getting her afloat. About ſeven the King and attendants were deſired to be preſent, and in a little time the veſſel was agreeably launched, to the general joy of every ſpectator. Never was there a more affectingly happy ſcene.——Every eye ſeemed to ſparkle with a luſtre borrowed for the occaſion.——Every countenance looked animating joy and heartfelt ſatisfaction; but few among them could utter their feelings: looks of congratulation circulated around, while every

every one shook his neighbour's hand with warmest fervour. Home, wives, parents, children, friends—all—all—seemed as within grasp—but description is unequal to this task. Let not however the behaviour of their Pelew friends be forgotten; in their joy which was also unbounded, real philanthropy was to be seen—They saw by this occurrence, those friends whom they valued about to leave them; those friends by whom they had been so much benefited, and from whom they had learned so much—but they saw them happy—they knew their whole comfort depended upon the success of this event, and therefore their benevolent hearts participated in the general joy.

After a very happy breakfast indeed, they proceeded to carry every thing aboard with all possible expedition, and in the afternoon, the flood tide coming in, in, the ship was hauled into the bason, a deep place of four or five fathom water; and in the course of the day, they got on board all the provisions, stores, &c. such only excepted as were to be given in presents to the King; and in the morning took on board their anchors, cables, and other necessaries, making bitts, and fitting a rail across the stern of the vessel.

Abba Thulle being now at the watering-place, sent for Captain Wilson to attend him; on whose arrival it was intimated to him, that the Rupacks had determined to invest him with the order of the BONE, and to create him a Rupack of the first rank; an' honour which Captain Wilson said, he considered in a very flattering point of view, and would receive with much joy. We shall mention this ceremony of investing with the BONE with some minuteness,

minuteness, as it is a distinction never obtained without the most unequivocal pretensions to merit, in the field, in the council, or in domestic life; and is esteemed a supreme felicity by the distinguished candidate. Some may be ready to smile, and ridicule the simplicity with which these children of nature, stamp this simple ceremony with so much consequence; but it were happy for European nations, if their marks of honour were as carefully conferred, and as surely indicated true merit in the possessor. The glaring ceremony with which the order of the Star and Garter or the dignity of a Peerage is conferred in Britain, does not convey more honourable sentiments of the distinguished object to the beholders, than this simple unadorned badge of honour does at Pelew; nay there, perhaps it may be justly said, the prospect of the

BONE,

Bone, excites more emulation, infpires greater courage, and more frequently promotes virtue and commands refpect, than the embroidered ftar or tinfelled ribbon do in Britain.

The King and Rupacks having retired to the fhade of fome trees, they enquired at Captain Wilfon of which arm he made moft general ufe, which having found to be the right, they took a circular bone, prepared for the purpofe, through which, with a good deal of trouble they compreffed his hand; after it had been fairly paffed over the joints of the hand, and fixed on the wrift, the King addreffed him nearly as follows. *" You are now invefted with our higheft mark of honour, and this Bone, the fignal of it, you will carefully keep as bright as poffible, rubbing it every day; this high mark of dignity muft always be valiantly defended, nor fuffered to be wrefted from*

from you but with your life." He was then complimented by his brother chiefs, on being admitted into their order; and the inferior natives, flocking round, shouted aloud to the *Englees Rupack.*

Monday the 10th, the old dwellings at the cove were cleared, and all the necessaries carried on board. The wondering natives so thronged the vessel, that they were obliged to complain to Raa Kook, who got orders from the King, that none but Rupacks, should go on board; but that the multitude might observe her, at some little distance in their canoes. As soon as the sails were bent, they took her to the west side of the island, and moored her in six fathom water. An immense concourse of natives followed in their canoes, hallooing and shooting in a most joyous manner. The King's two brothers accompanied

companied them, who repeatedly called to their people to be minute in obferving the management of the fhip; as if, at fome period, they expected to have one of their own. After this the Captain went on fhore to the King, who was waiting for him at Oroolong. Abba Thulle now refumed the fubject of fending two of his people to England: he told the Captain, by means of the interpreter, that he had the happinefs of being much refpected by all his fubjects, not only as being fuperior in rank, but in mental capacity; but notwithftanding, he had often felt his own infignificance, in feeing the meaneft Englifhman, exercife talents of which he had no conception; and had therefore refolved to part with his youngeft fon, Lee Boo, who fhould, in company with one of the Malays, as an attendant, be entrufted to Captain Wilfon's

Wilfon's care, that he might be inftructed in fuch fciences as would tend, on his return, to advance the profperity of the people, and reflect honour on the royal family. This youth, he faid, was of a mild, pliable temper, and an enemy to every kind of vice; he was under the care of an old man who lived at fome diftance; but had orders to be at Oroolong in the morning. Captain Wilfon replied, that this mark of his regard and efteem affected him deeply, and he fhould ever think of his confidence with pride; he affured Abba Thulle, that any perfon belonging to Pelew would meet with attention from him, but the fon of the man to whom he had been fo much indebted, he held himfelf engaged by every tie, to treat with the fame tendernefs as his own fon.

Nor were thefe the only perfons who wifhed to accompany the Englifh; for fome time an uncommon gloom had obfcured Raa Kook's chearful countenance, which feemed to increafe as the hour approached when he was to part with his beloved *Englees*. It was afterwards difcovered, that fo great was his attachment to the Englifh, that he had afked permiffion of his brother, the King, to accompany them home; this however could not be granted, as Abba Thulle reminded him, he was next to him in command, and in cafe of death his fucceffor; that therefore it would be exceedingly imprudent in him to attempt it. A nephew of the King's at fame time made application to the fame purpofe; it was his brother who was flain at Artingall as before-mentioned. This young man urged his fuit to Captain Wilfon with great importunity, but

but the Captain declined till he had obtained permission from his uncle. Captain Wilson conversed Abba Thulle on the subject; he replied, that the young man was unworthy of his protection, having rendered himself disagreeable throughout the island. At the same time, the young man appearing to plead his own cause, the King sternly addressed him nearly to the following purport. *" You are undutiful to your aged mother, and though you have deserving wives, you use them ill. Your vile conduct has been publicly exposed, and now you would fly from the resulting shame! Remain where you are; be ashamed of your conduct, and reform."*

Abba Thulle now modestly hinted a request to Captain Wilson, that before he set sail, he would sail round in his new vessel to Pelew; he particularly mentioned that many aged people there had

had never yet seen their vessel, &c. and were very anxious to do so; he said it would not detain them long, and be very agreeable to all his people. The Captain at once conceived that this proposal would be very disagreeable to his men, and perhaps revive their former suspicions, he therefore suggested such objections as satisfied the King that the step would be improper.

The weather and wind appearing favorable, the Captain informed Abba Thulle, that they purposed sailing the next day. This very much distressed him; we have formerly mentioned the embassy which had been sent him at Pethoull, in answer to which he had sent word that the day following that now mentioned by the Captain, was the day the English meant to sail, in consequence of which all the neighbouring Rupacks were to come to O-roolong

roolong the next night to furnish them with provisions, and bid them farewel. This information determined the Captain still more to set sail in the forenoon, as the number of canoes to be expected, would greatly incommode them, he therefore apologised in the best manner he could to the King, who appeared greatly disappointed. He then begged that the Captain and officers would dine with him and his brothers on shore. With this they cheerfully complied, and after dinner Arra Kooker so pathetically begged for the favourite dog, of which he had become excessively fond, that they could not resist his solicitations, though it would prove a particular regret among the sailors. But the general's intention was far otherwise employed, he was already building a ship in imagination; and to realize his design, wished them to leave their

launching

launching ways, saying he would go to work on the same place. The King had laughed at the insignificance of Arra Kooker's request of the dog; but the subject of ship-building caught his most serious attention; it was of national importance, and of course demanded the patronage of a good prince. In the midst of their discourse a battle on board the ship between two sailors, called for the presence of the Captain: the damage proved no greater than a bloody nose, which being settled, Mr Wilson again returned. When the circumstance was explained to the King, he observed that there were no doubt bad men in all countries. The English permission was asked, and obtained, to hoist an English pendant on a tree near the cove, with an inscription as follows, on copper, to be placed on another tree adjacent:

THE

THE HONOURABLE
ENGLISH EAST INDIA COMPANY'S SHIP
THE ANTELOPE,
HENRY WILSON, COMMANDER,
Was loſt upon the Reef north of this Iſland,
In the Night between the 9th and 10th of Auguſt;
Who here built a Veſſel,
And ſailed from hence the 12th of November 1783.

Captain Wilſon explained the purport of this inſcription to Abba Thulle who was greatly pleaſed with it; and havng explained it to his people, he aſſured the Engliſh, that it ſhould carefully ſtand there in remembrance of their viſitors.

The converſation this day was principally confined to the approaching ſeparation. "When you are gone," ſaid the King, "I much dread that the Artingalls will redouble their attempts againſt me; and deprived of your aid, I ſhall probably feel the effects of that animoſity they have always had towards my people; and having no more the English to support me, I will not be match for them, unless you leave the few musquets you promised me." The Captain was quite satisfied to comply with the request immediately; but most of the officers who still had apprehensions, were unwilling to give up the arms till the last moment; that unlucky suspicion which had so ungenerously taken possession of them, had been so rivetted in their minds, that it was not easily dislodged. It is necessary however, not to condemn our countrymen too hastily; they had been accustomed to see roguery so generally, and so scientifically practised, that distrust and suspicion are naturally instilled among the first principles of education; and it was not easy for them to conceive, that the same species should be so very different even at opposite quarters

quarters of the globe; but here they saw, the open undisguised actions of nature, knowing no deceit and dreading none.

Abba Thulle was too quick sighted not to observe their distrust; and it is not easy to express the agitation which laboured in his breast, on finding that doubts were harboured of his sincerity. " Why," said he, " should you distrust me? I never refused you my confidence. If my intentions had been hostile, you would have known it long ago, being entirely in my power: but on the contrary, you have had my utmost assistance; and yet at the very last you suspect me of bad designs!" The earnestness of his manner, spoke his feelings more than his words; nor need it be doubted, that a little recollection brought the blush into the countenances of those whom he addressed. The man

man who had uniformly behaved with such disinterested, unsuspecting benevolence—the man who freely committed his own son to their care, to be doubted within a few hours of their parting, was a stab, which the sensibility of Abba Thulle could not support: the severity and truth of his reproach, and the noble dignity with which he supported himself, brought the daring thought of butchering him and his brothers to view, and gave a most captivating picture of the mild, yet forcible triumph of virtue. They found themselves guilty, and saw evidently that virtue will flourish in whatever soil she is implanted. Without further hesitation, they sent on board for all the arms that could be spared; and on the boat's return presented him with five musquets, five cutlasses, more than half a barrel of gun powder and

flints

flints and ball in proportion. Once more harmony was reſtored, and the generous Abba Thulle forgot, or ſeemed to forget, the cruelty of their ſuſpicions.

The King's ſecond ſon, Lee Boo, arrived in the evening from Pelew, under the care of his elder brother: Abba Thulle preſented him to the Captain, and then to the officers; he advanced in ſo eaſy and polite a manner, having much good humour and forcible expreſſion in his aſpect, that every one was prepoſſeſſed in his favour. As it was now getting dark, the officers went on board, leaving the Captain behind, at the King's requeſt. Next day Mr Wilſon informed them, that neither he, the King, nor the Rupacks, enjoyed much reſt; the affectionate father employing the moments in giving advice to his ſon, and in recommending him to the care

care of the Captain; not, however, from the fmalleft fear that he would be ill-treated: " I would wifh you," faid he, " to fhew my fon every thing that is ufeful, and make him an Englifhman. The fine things he will fee may probably induce him to flip away from you, in fearch of lefs confined gratification; but I beg that you will contrive to calm and fubdue the rafhnefs and impetuofity of his youth. I well know from the different countries he muft pafs through, that he will be liable to dangers, and even to difeafes that we never heard of, which may kill him; but I alfo know, that death is the common lot; and whether he dies with you, or at Pelew, is of no moment. I know you are a man of humanity; and am therefore confident that if my fon be fick, you will look on him with kindnefs. But fhould that

that happen, which your utmost care cannot prevent, let it not deter you or your brother, or any of your countrymen, from returning; for I shall rejoice at the sight." The Captain assured the King, that he might rest satisfied of the care and affection with which his son would be treated.—Before Mr Wilson came on board, he admonished Blanchard (the man who had resolved to renounce his country) as to his conduct among the natives; he desired him to be watchful of the arms and ammunition that would be left behind, that they might defend themselves from their enemies; he begged him not to go naked, like the natives, as it might lessen his importance with them as an Englishman, and countenance an evident indecency; and that he might have no excuse from the want of cloaths, all that could be spared was left him; in order

order, alfo, that if he accepted the King's offer of wives, he might be enabled to drefs them fomewhat after his own cuftom. The Captain did not forget to enforce on him the abfolute neceffity of continuing his religious duties, and to be particular in keeping Sunday on Sabbath. After this, he was requefted to afk any favour that might tend to his future comfort; on which he begged to have one of the fhip's compaffes, and the mafts, fails, and oars belonging to the pinnace, which alfo was intended to be left behind.

Wednefday morning early, an Englifh jack was hoifted at the maft-head of the Oroolong, and a fwivel fired as a fignal for failing; which being explained to the King, he ordered all the provifions on board, which he had brought for our voyage. A great number of canoes furrounded the veffel, loaded

loaded with presents, so that it was with difficulty they could avoid being overstocked. When just ready for sea, a boat was sent on shore for the Captain, who then took Blanchard and the men of the boat, into a temporary hut that had been erected; and, making them kneel, offered up thanksgivings to that Power who had supported their fainting spirits through so many hazards and toils, and had at last opened to them the door of deliverance. He repeated his advice to Blanchard, earnestly begging him not to forget his religion.

When Lee Boo came to the watering-place, there were sent with him three or four dozen of a very fine fruit, similar to the English apple; it is of a fine crimson colour and oblong shape: this is a very rare fruit at Pelew, though there are plenty of them in the different

ent South Sea Iſlands. One of them was given to every officer, and the remainder kept for Lee Boo.

About eight o'clock, the Captain went on board, attended by Abba Thulle, Lee Boo, the Rupacks and Blanchard. It being doubtful as the veſſel was heavily laden with proviſion, whether, ſhe would be able to get over the reef, it was reſolved to land the two ſix pounders and leave the jolly boat behind, as they had nothing wherewith to repair her, and ſhe was almoſt worn out; in her room Abba Thulle was at great pains to procure them a proper canoe.

Captain Wilſon had recommended Mr Sharp the ſurgeon to Lee Boo as his *Sucalic* or friend, and the young man ſtuck by him, with the greateſt attention, attending him to whatever part of the veſſel he moved, as his Mentor

Mentor on all occasions. Blanchard now got into his pinnace, in order to take the veſſel in tow, and parted from his old ſhip-mates with as much compoſure, as if they were to meet again after a ſhort abſence; he ſhook hands with them with the ſame indifference, as if they were about to ſail down the Thames on a coaſting voyage—a ſtriking contraſt to what followed!

The veſſel now proceeded towards the reef, deeply laden with Abba Thulle's bounty to a degree of ſuperfluity, and ſurrounded by great numbers of the natives in their canoes, who had every man brought his preſent, for their good friends the *Englees*—What a luxurious ſight to a feeling heart!——— There was no room for them, yet every one exclaimed, *only this from me, only this from me*, and if refuſed, they repeated their requeſts with ſupplicating countenances

countenances and tears in their eyes. Indeed their generofity and affection were fo urgent, that a few trifles from the neareft of them were accepted; while the others, unable to bear the feeming neglect, paddled a-head, and put their little prefents in the pinnace, not knowing that fhe was again to return to the fhore. Several canoes went before the veffel to point out the fafeft track; and others were waiting at the reef, to fhew them the deepeft water: from all thefe precautions, which were directed by the King, the reef was fortunately cleared without any accident. The King now came along-fide, and gave Lee Boo his bleffing, which the youth received with great refpect and tendernefs: he next embraced the Captain, in much apparent diftrefs, and then cordially fhook hands with all the officers, crying——
" You are happy becaufe you are going home,

home, and I am happy becauſe you are
ſo; but ſtill very unhappy at your going away." Once more renewing his
aſſurances of regard and good will, he
left the veſſel, and went into his canoe.
The natives who were to return with
the King, looked up to the veſſel eagerly,
with the moſt expreſſive countenances,
and half diſſolved in tears. This proof
of delicate ſenſibility, and of proved
affection, operated ſo ſtrongly on the
feelings of all aboard, that it was with
much difficulty they ſummoned reſolution enough to give three cheers at
their final departure. Raa Kook remained, with a few of his attendants,
to ſee them out of danger beyond the
reef; but was ſo highly dejected, that
the veſſel had gone a great way before
he thought of ſummoning his canoes
to return. As he had been their firſt
friend, the Captain gave him a brace

of piſtols, and a cartouch-box with cartridges: and the moment of ſeparation being now come, he appeared ſo much affected, that it was ſome time before he could ſpeak; pointing to his heart, he ſaid it was there he felt the pain of bidding them adieu. He endeavoured to converſe with Lee Boo, his nephew; but being unable to proceed, he precipitately went into the boat, and giving them an expreſſive glance, as if his mind was convulſed, he inſtantly dropped a-ſtern; and thus terminated our connection with the natives of Pelew, after a reſidence among them from Sunday the 10th of Auguſt 1783, to Wedneſday the 12th of November following.

It may not be unſeaſonable, while the amiable behaviour of theſe two reſpectable characters, Abba Thulle and Raa Kook, is freſh in the recollection

lection of the reader, to make a few general observations on their different characters.

Never was a prince more formed to attract and retain the love and admiration of his subjects than Abba Thulle; his appearance majestic, he commanded with authority; while his affability and easy access, rendered him a semideity, to all his subjects. In one of his councils, there was as much (we had almost said more) respect paid to his naked unadorned person, as to a European potentate, amidst all his trappings and pageantry, from the surrounding sycophants. His nice honour and quick feelings were very discernible on many occasions; never was there a reproof more delicate and yet more poignant, than what he gave the English on occasion of the late affair with the musquets. He was far from one

one of those harmless *nothings*, who hurt nobody because they have not a sense of injuries; while the warmth and sensibility of his heart won the love of all around him, his dignity of manner, and propriety of conduct taught them to approach him with respect. He possessed a contemplating mind, and few objects came within his observation, without being attentively considered. The prosperity of his subjects, was the principal object with him. It was this led him to part with his son Lee Boo, whom he tenderly loved——
For this he was at so much pains in examining every thing about the English, that might be serviceable to his people—in fine, his whole attention was engaged in forming and executing plans for the good of the nation and individuals. In domestic life he shone remarkably, and took a particular charge

charge of all his own relations; the misbehaviour of his nephew, at which we have already hinted, seemed to give him the greatest pain; while as the husband and parent, his heart seemed awake to every finer feeling which adorns humanity.

Accident only has made him acquainted with a few of the rest of mankind; and that accident he considered as the happiest of his life; we may perhaps never hear of him again, but judging from what is already known, he may justly be considered as one of the best of men and of kings.

His brother Raa Kook was a prince of so universally engaging demeanor, and whose every action expressed something so truly valuable, that Englishmen or natives equally admired him. He was so much a friend to the English,

lish, that it may be suspected their account of him is partial; therefore little shall here be said, and that little not exaggerated.

His natural temper was cheerful and pleasant, though without that mimicry and humour for which his brother Arra Kooker was remarkable; at the same time he was far from averse to a good hearty laugh when a proper occasion offered. As commander in chief, he was beloved by them all: he dispensed his orders calmly and smoothly, but would not tolerate neglect. No man better understood the necessity of strict discipline; so that while he encouraged his inferiors to use all becoming freedom with him, he kept them at that *proper* distance, which is the true key to cheerful obedience. In principles of honour, he was by no means inferior to his brother; and
not

not only wifhed that the Englifh fhould hold *him* in an honourable point of view, but all the nation; thus it was, that he could not bear the leaft idea of pilfering among them, for, as formerly mentioned, if any thing was amiffing, Raa Kook foon difcovered and punifhed the delinquent. One day, a chief Rupack fought a cutlafs from Captain Wilfon in his hearing; the frown inftantly appeared, nor would he fuffer it to be given him. He was exceedingly delicate in receiving favours himfelf; and though from his particular difpofition in enquiring after caufes and effects, many things about the Englifh were very highly prized by him, he was particularly attentive, that nothing fhould betray any defire for what he thought might not be proper to be given. The reader has already feen, his agreeable deportment

in

in his family; even to a degree which many in this age of diffipation and ftoicifm might reckon filly; but let it be noticed, that though the finer feelings fhone in the natives of Pelew, to a length many in Britain would call effeminate—yet in fatigue, pain, diftrefs and death, they appeared as heroes indeed.

Before we proceed to mention the future fortunes of our navigators, the following chapter is introduced to mention fuch obfervations on the manners and cuftoms of thefe amiable people, as are thought interefting.

CHAP.

Chap. VI.

General Defcription of the Iflands—Productions—Natives—Drefs—Difpofitions—Manners—Religion—Marriages—Cuftoms—General Character—Government—Precedency, &c.

AS the Antelope was not a veffel fitted out for difcovery, and furnifhed with fcientific gentlemen, qualified for making many philofophic obfervations, the naturalift, or philofopher, muft wait the iffue of more particular difcoveries and enquiries. Men diftreffed with the dread of perpetual exile, and whofe attention was almoft wholly occupied about their deliverance, were not the perfons for tracing nature accurately in her various appearances and effects.

The Pelew Iflands, or as fome call them the Palos Iflands, are fituated between 130° and 136° of eaft longitude from London, and 5° and 9° north latitude. They are long and narrow, lying in north and east direction. They are plentifully covered with wood of various kinds; such as the *Cabbage Tree*, *Ebony*, and a species of the *Manchineel*, the sap of which, when it touches the skin, occasions an immediate swelling and blistering; this was the tree, which they considered as unlucky. But their three most remarkable trees, we in Britain are totally strangers to; one is a very pretty tree, and upon boring a hole in it, a thick substance like cream distills from it: another is very like a cherry tree in its manner of branching; it has a very thin cover, which is not properly a bark, being as close in the texture as the inner wood, which is very hard; none of the English tools could stand to work it: in colour it is very like, though still prettier than mahogany,

hogany: the laſt is like an almond tree, the natives call it carambolla. Beetle-nuts, yams, cocoa-nuts and bread-fruit, are their ſtaple articles of livelihood, about which they are principally concerned; and a few oranges, lemons and the jamboo apple (thoſe brought to Lee Boo on his departure) are their delicates. They have no grain. The iſlands are in general well cultivated, as the natives ſpare no pains; all their labour conſiſts in fiſhing and the cultivation of their grounds. Every man had his own piece of ground ſo long as he inclined to dwell there; but if he left it for another, it returned to the King, as chief proprietor, who beſtowed it on the next that applied for it. One thing was very diſcernible, that every man had his own canoe, which he kept ſacred.

It has already been mentioned, that there are no quadrupeds on the island, except rats. Birds of different kinds were observed flying about, some of them very beautiful, but the greater part of them are those which are known by the name of tropic birds. Whether from their peculiar kinds, or the echoing in the wood, is not easily determined, but the English were ready to think their notes had a very peculiar melody; one in particular, was uncommonly sweet, but though the sound appeared quite at hand, none of the birds could be noticed. But we must not omit to mention, that the English have probably taught them a lesson which may be of great service to them; the islands abounded with common cocks and hens, which the natives considered as a very useless animal, and therefore took no pains about them, but

but left them to wander wild through the woods; at times they would have eaten their eggs, provided they were to their taſte, that is, not freſh or lately laid; but if containing an imperfect chicken, they were delicious. They were now however taught to eat the fleſh of the fowls, which they ſoon found to be a very palatable food.

Few parts of the globe are ſo well ſupplied with fiſh of all kinds, particularly mullets, crabs, oyſters, muſcles, &c. but the fiſh moſt eſteemed among them is the ſhark, the greater part of which they reckon delicious. Several kinds of ſhell fiſh, they eat quite raw, in preference to dreſſed. They have few freſh-water fiſh, as there are no rivers on the iſlands, only a few pools and ſmall ſprings, &c. Their method of preſerving their fiſh, has been already noticed; they have no ſalt, and have little

little conception of sauce or seasoning to any thing they eat. Sometimes they boiled both fish and vegetables in saltwater, but this was no improvement; but when they eat any thing raw, they squeeze a little orange or lemon juice upon it.

They get up betimes in the morning, and their first work is to bathe. There are particular places appointed for this; and a man dares not approach the women's bathing places, without previously giving a particular halloo, of which, if no notice is taken, he may proceed, but if they halloo in return, he must immediately retire. They breakfast about eight, and proceed to public business or any other employment till noon, when they dine; they sup about sun-set, and very soon after retire to rest.

The

The reader will have obferved frequent mention is made of fweetmeats in this narrative, a more particular account of which may be proper. They had various forts; one was prepared by fcraping the kernel of the cocoanut into a pulp, and then mixing it up with orange juice and fweet drink. This fweet drink is a compofition of the juice of fweet canes, which the ifland produces plentifully. This mixture they generally fimmered over a flow fire, which when warm they made up into lumps; it foon turned fo hard that a knife would fcarce cut it. This the Englifh called *Choak-dog*, but the natives called it Woolell. Another fort is made up of the fruit of the tree juft mentioned like the almond tree; and on one occafion they prefented Captain Wilfon with fome liquid fweet meats,

meats, which they prepare from a root somewhat similar to our turnips.

The natives are in general stout, well made and athletic; many of them appeared to be uncommonly strong; they are in general about the middle size, and universally of one tinge as to colour, not wholly black, but a very deep copper colour. The men have their left ear bored, and the women both; they wore a particular leaf, and at times an ornament of shell in the perforated ear. Their noses are also ornamented, by a flower or sweet shrub, stuck through the cartilage between the nostrils. This custom is not peculiar to Pelew, but is found in many eastern nations, and probably proceeds from their great desire for sweet scents; and though at first it appeared rather disagreeable, from want of use, it is certainly a more pleasant and becoming

refreshment

refreshment to the nose, than the use of tobacco either by snuffing or plugging. Their teeth we have already mentioned are died black; but the English could never learn the method it was done, nor more about it than that it was accomplished by means of some herbs when young, and the operation was very painful. The tatooing the body is also done in youth, though not altogether in childhood. The only appearance of any thing like dress among these natives is in the female sex, who in general wear a piece of mat, or the husks of cocoa-nut died, about nine or ten inches deep, round their waist; some of these aprons are very neatly made, and ornamented with beads, &c. Abba Thulle's daughter Erre Bess, gave Henry Wilson a present of a very neat one to carry to his little sister.

It

It has in general been granted, that mankind however ignorant and favage, are ftill poffeffed of confcience, and the internal knowledge of a certain fomething, their fuperior, to which they are accountable; nor has any clafs of men yet been difcovered who have not fome outward rite or ceremony whereby this knowledge is expreffed. However fuperftitious, enthufiaftical, or foolifh their different modes of worfhip may appear; to thofe who have been bleft with revelation, nay, however, much we may be puzzled to inveftigate a caufe, to which the fingularity of fome of their religious rites are to be attributed, yet ftill the exiftence of one Great Firft Caufe or ruling Deity has been acknowledged, not in word only, but by fome outward ceremony or rite. To deny that any fuch cuftom exifts at Pelew, and yet that they ac-
knowledge

knowledge a Superior Power, may be confidered rafh, yet, from the moft attentive obfervations and enquiries the Englifh could make, they have reafon to believe that is the cafe. Neither place, time nor circumftance, could be obferved as pointing to any worfhip or religious rite; nor could the Englifh collect any thing from their converfation, though particularly queftioned on the fubject, from which they could difcover their ideas concerning the God of nature. The moft probable conjecture is, that the inward monitor, at which we have juft been hinting, leads them to think of fome directing chance, good and bad, without any percife idea further.

In order that the reader may fomehow be enabled to judge for himfelf in this particular, he may recollect the following circumftances already mentioned

oned—The *unlucky* wood which Abba Thulle mentioned to Captain Wilson—Raa Kook's behaviour in the old woman's house with the nuts, &c. after his son's funeral—and the muttering which took place on several occasions, mentioned through the preceding sheets. It was very clear, that they had some strong fixed idea of *Divination;* when Lee Boo set out to sea, he was for several days uncommonly sick; and he then told Mr Sharp, he was sure his father and friends were very sorry for him, for they knew what he underwent. He was prepossessed with the same idea when dying, as we will soon have occasion to mention. Indeed, on one occasion, while in Britain, he seemed to intimate that they understood the spirit existed even after death; as upon occasion of Captain Wilson's informing him the intention of going to to church, being to reform men's lives, and that they might go to heaven, he replied, that at Pelew, bad men stay on earth, and good men grow very beautiful and ascend into the sky. When Mr Barker fell from the side of the vessel, the natives said it was owing to the *unlucky wood* being in the vessel; and upon several other occasions seemed to hint at the effects of a superior power. One particular mode of divination was observed, and considered to be peculiar to the King, as none but he used it. They have a plant, not unlike our bulrush, by splitting the leaves of which and applying to the middle finger, he judged of the success of any occurrence of moment; before the first expedition to Artingall, it was noticed that the answer was very favourable, but when about to set sail on the second, the oracle did not appear altogether fo agreeable; Abba Thulle therefore would not fuffer them to enter their canoes, until he had twifted his leaves, till as he thought they appeared more favourable. On this fubject we fhall only further add, that the refpectful attentive filence of the natives, while the Englifh were at worfhip, feemed to indicate, that although they knew nothing of any religious forms of worfhip, yet they were not infenfible of the exiftence of one *Great Supreme Caufe,* who rewarded and punifhed according as deeds fhould merit; and hence that ftrong fenfe of propriety, juftice and delicacy, which produced among them the ftrictest morality.

The general character of thefe natives of Pelew is now pretty well imprinted on the reader's mind, a very few additional obfervations are therefore

fore neceffary. Humanity is the prominent feature in the picture; the Englifh were caft upon their territories, in a ftate the moft helplefs that can well be conceived; twenty feven men, without even common neceffaries of life, entirely dependent on their bounty; fed, fupported, affifted in their labours, and every thing done for them that was in their power. Let us only for a moment confider the hourly bounty which was poured in upon them, not of their ufelefs provender, but, as the Englifh had many occafions to obferve, their beft provifions were given to their ftrangers, while many perhaps were fcanty enough at home. Only recollect the parting fcene—fee the crowding canoes holding out prefents, not the diftant effects of complaifance, but the warm effufions of philanthropy!— Could oftentation, pride, or the hope of retribution influence them? by no means—it was kindnefs to men they never expected to fee again.

Their native politenefs was conftantly obfervable; poffeffing a degree of curiofity, beyond any of the South-fea natives, they never knowingly intruded when it was inconvenient; in them it was evident that *good manners* are the natural refult of *good fenfe*. The attention paid by the men of Pelew to their wives, was very uncommon in moft parts of the world; and even a Britifh hufband might at times get a leffon. Their marriages feemed to confift in a ferious folemn contract without any formal ceremony, but they are ftrictly faithful to one another; and the utmoft decency of behaviour is uniformly fupported. A hufband never fleeps with his wife when pregnant, but during that period the greateft attention

tion is paid to her, in order that she may be kept easy. A plurality of wives is allowed, though they generally confine themselves to two, a Rupack three, and the King five: they name the children soon after born, without any ceremony. One of Abba Thulle's wives bore him a son while the English were there, which he named *Captain*, to the memory of Captain Wilson. They are far from being naturally lascivious, and the utmost decency is preserved among the natives; one of the sailors endeavoured to pay his addresses to a female, but was rebuffed in a manner that prevented any further attempts.

They are in general an active, laborious set of people, possessing the greatest resolution in cases of danger, patience under misfortunes, and resignation at their death. Except a few Rupacks there was little subordination of rank,

rank, (and of that we will speak presently) consequently their employments were pretty much the same; fencing their plantations, planting their yams, making hatchets, building houses and canoes, mending and preparing fishing tackle, forming darts and warlike weapons with domestic utensils, and burning chinam, may be said to comprise the whole round of their employments. Those who had a particular turn for mechanical operations or any uncommon pieces of work, they called *Tackelbys*; it was to them the King so often gave particular orders to observe the building of the Schooner. Idleness was tolerated in none; the women were as laborious as the men, and the King and Rupacks were as much employed as any. Abba Thulle was the best maker of hatchets in the island; and generally laboured at them when

when disengaged from affairs of state; they had no idea of unemployed time, and therefore it is, that without the proper tools for finishing a fine piece of work, practice had taught them, even with their coarse implements, to execute, what a British Artist, could not have conceived practicable. Their mats, baskets and ornaments are so curiously wrought, that when their simple tools are considered, the ingenuity is more to be admired, than much superior productions executed under the advantages which European mechanics enjoy.

That equality of station which appeared evidently among them, and ignorance of those luxuries which civilization intruduces, proved no inconsiderable source of happiness to them; the one prevented that ambition which is often so destructive to society, and the other those cares which
affluence

affluence awakens. In all the connection which the English had with them, robbery or rapine were never named among them; nature it is true, allowed them little, but that little they enjoyed with content. Human nature here shone in most amiable colours; men appeared as brethren; uninformed and unenlightened, they grasped at nothing more than competency and health; linked together as in one common cause, they mutually supported each other; courteous, affable, gentle and humane, their little state was cemented in bonds of harmony;——but a short account of their government may be proper.

Abba Thulle, the King was the chief person in the state, and all the homage of royalty was accordingly paid to his person. He was supreme in the greater part of the islands which came within

thin the observation of the English; but Artingall, Pelelew, Emungs and Emellegree appeared to be independent, tho' from any thing that could be observed or learned, their form of Government was similar. The general mode of making obeisance to the King, was by putting their hands behind them and bowing towards the ground; and this custom prevailed not only when passing him in the streets or fields, but when they passed the house in which they supposed him to be. His carriage and demeanour was stately and dignified, and he supported his station very becomingly. He devoted the forenoon to public business, and decided every matter of state by a council of Rupacks. They assembled in a square pavement in the open air, the King being remarkable from being placed in the centre, on a stone of larger size than those of the

the Rupacks. They seemed to deliver their minds with freedom, as matters occurred; and the assembly was dissolved by the King rising up. The afternoon was devoted to receiving petitions, hearing requests, and deciding controversies; these it may easily be supposed seldom occurred, for as their property was small and of little value; and as there were no lawyers nor their emissaries to foment disputes, the proper barriers of right and wrong were easily defined. Wrangles and fighting seldom happened, for even a dispute between children was checked by a severe frown, and their impetuosity bridled. When any real injury was done by any one to his neighbour, it was a pleasing sight to see how justice was administered; their laws were the simple dictates of conscience as to right and wrong
between

between man and man; no rhetoric or enticing words of wifdom were employed to mafk vice under the cloak of virtue; none of thefe fubterfuges could be employed whereby fraud and oppreffion could be fcreened; oaths were unknown, and the fimple dictates of truth directed the judge; nor were there any punifhment of a corporeal kind; being convicted of injuring a neighbour was to them more galling and difgraceful, than any pillory yet invented by Britifh ingenuity.

Meffages were tranfmitted to the King with great ceremony; the meffenger never was admitted into the prefence, but delivered it to an inferior Rupack, who delivered the meffage to the King, and brought his anfwer.

The General was next in authority to the King, and acted for him in his abfence; he fummoned the Rupacks to to attend when needed, and had the chief command of all the forces; tho' it was obferved, that in actual engagement, when the King was prefent he himfelf officiated. The General fucceeded the King in cafe of his death, and on his demife, Arra Kooker; when the fovereignty would again revert to Abba Thulle's eldeft fon, then Lee Boo and fo on. The King had always an attendant, who though not fo high in office as the General, was more conftantly about his perfon. He was confidered as the principal Minifter, and a man of judgment; he never bore arms, nor went on the warlike expeditions. It was remarked that he had only one wife, and never invited any of the Englifh to his houfe.

The *Rupacks* were very numerous, and confidered in the fame light as the nobility are in Britain. They were of
different

different orders, diftinguifhable by the fize and quality of the Bone, of which Captain Wilfon belonged to the higheft rank. They all attended the King on command, every one bringing with him, a certain number of dependants, with their canoes, fpears and darts. The reader will be apt here to trace a fimilitude to the feudal fyftem; but as the knowledge the Englifh acquired of thefe matters was very fuperficial, nothing very conclufive can be afcertained. Thefe and many other matters muft be left for time to develope; all that can be faid further at prefent is, that whatever was their precife mode of government, it was wonderfully adapted for the people. All the iflands appeared populous, but the number is not eafily conjectured. There were four thoufand active men in the expedition againft Pelelew, and it was evident, many more were left at home, not being needed.

The method they took for building houses, was very ingenious. They raised them three feet from the ground, in order to prevent damp; this space they filled up with solid stone and overlaid with thick plank as a floor. The walls were built of wood, very closely interwoven with bamboos and palm leaves, so that no cold nor wet could possibly come through; the roof was pointed in the same manner as village houses are in this country. Their windows come down on a level with the floor, answering for doors also; and have a sort of shutters, which they fill up the chasm with, when necessary; their fires are kindled in the centre of the room (for all the house is in one room) the fire place being sunk lower than the floor, with no timber below it, the whole

whole fpace being filled up with fmall ftones, &c. Their houfes for public meetings are about 70 feet long, but the common dwelling houfes do not exceed forty.

Their fpears have already been mentioned, it is thought only neceffary to add concerning them, that they were barbed tranfverfely, fo that if once they were ftuck in the flefh, it was impoffible to extract them. One of their moft effective weapons in war is the dart and fling; the darts are thrown by means of an inftrument, not unlike what is called a crofs-bow in this country; upon a ftrong ftraight piece of wood the dart is laid, and on one end of the wood is fixed an elaftic piece of bamboo, by compreffing which with greater or lefs force, they throw the dart to the diftance of fifty or fixty feet as they incline; it is aftonifhing how exactly they can direct this weapon, and the distance at which it proves mortal. They have daggers about thirteen inches in length, which are made of bones of fish; and some of the Rupacks had also a kind of sword, made of very hard wood and inlaid with shells.

Their canoes are admirable workmanship; they are made of the trunks of large trees, in the same manner with those throughout the South-seas, but with surprising neatness; they ornament them with shells and paint them red. It has already been mentioned, that the natives painted the Schooner for the English, and as their method is peculiar, it may be proper to mention it. They take the red ochre and crumble it down among water, then soak it for some time over a slow fire; when it is brought to the necessary consistence,

confiftence, they rub it on the wood, while warm; when dry, they varnifh it with cocoa-nut oil, which gives it a polifh that no water can efface. Their canoes are of different fizes, but the largeft will not carry more than thirty people; the common ones, from five to ten. They ufe fails made of matting, which are by no means capable to encounter a rough fea, they therefore keep as near the fhore as poffible. The natives row with great ability, infomuch, that the canoes made purpofely for fwift failing, feem fcarcely to touch the water, moving with a velocity unknown to our boatmen; their dexterity in handling the oars is admirable, when on vifits of ceremony, they flourifh their paddles with great addrefs and exact order. Their domeftic implements are few in number and very fimple; they are the evident productions

tions of neceffity, well calculated to anfwer the ends intended, without much ornament. Their little bafkets which they always carry about with them, are among their neateft pieces of workmanfhip; in them they carry their nuts, knife, ftring, and any other little article they may need for the work about which they are employed. They are alfo hung up around the walls of their houfes for ornament and ufe. Their knives are made of fhells which they fharpen to fuch an edge as fully anfwers all common purpofes.

Their fifhing hooks were ingenioufly formed of tortoife-fhell, and their combs of the orange tree; the mats on which they flept, and thofe with which they were covered when afleep, were formed of the hufks of cocoa-nuts. They have a number of veffels made of earthen ware, for purpofes

fes of cookery; they ftand the heat exceedingly well, if heated by degrees, of which the natives are very careful. Their ftrings, cords, and fifhing-nets are all manufactured from the hufks of cocoa-nuts. Their drinking cups are made of cocoa-fhells, which they polifh with great art. But the hatchets are the moft uncommon weapons; the blade is made of a very ftrong fpecies of cockle, which they call *Kima Cockle*, ground to a fharp edge; they were very anxious however for iron to fubftitute in its room. Some of their hatchets were made with moveable heads, with which they could anfwer the principal purpofes of an adze; fo that though their tools were not remarkable for beauty or convenience, yet the Englifh were often furprifed with what facility they cut down the largeft trees with them.

Their

Their articles for ornament were far from numerous; the King had a very fine tureen, somewhat in shape of a bird, and finely ornamented with various devices, very neatly cut out upon it, this he made a present of to Captain Wilson; it held about 36 English quarts. The tortoise-shell they wrought into various little dishes, spoons, trays and other vessels; the shell they have in these islands is of a very beautiful kind, but their manner of working it, the English could never get an opportunity of observing.

The torches they use have been frequently mentioned; they seemed to be a rosin mixed with small pieces of a particular species of wood, which burns well; they have a clear light, and an agreeable smell.

Having now recapitulated such observations as the English had opportunity

nity to make during their fhort refidence among them, it may not be improper to glance a little at fuch parts of their conduct as feem to difagree with that ftrong humanity and urbanity which have been uniformly reprefented as diftinguifhing characteriftics in thefe natives of Pelew. Several inftances have been given of their noble principles not being confined to their intercourfe with one another, but that to their enemies, they behave with a degree of generofity totally unknown among nine tenths of mankind. Where is the nation that fcorns to attack their enemy unfeen or by night? What people fend embaffies to herald their approach? But it may be afked, how can it be accounted for, that thefe friends of humanity, fhould fo wantonly take away the lives of their fellow creatures, when captivated in battle: An anfwer

to

to this has already been attempted; the practice has but very lately taken place, and is the result of what they imagined to be *political necessity.* They considered a captive as a most dangerous person among them; no prisons, nor public works to employ them in, so that had they not put them to death, they would have been very troublesome companions.

The reader may also have noticed in the course of the narration, an inclination to pilfer, perhaps inconsistent with that character of integrity we have been just mentioning. The circumstances however should be carefully weighed; a nail, or bit of old iron, was to them a precious jewel; and the commonalty had almost no opportunity of getting any, except in the way of picking them up when they accidentally fell in their way; they very seldom
took

took any thing of confequence, except when the temptation was fo peculiarly fingular that even the rigidly virtuous would eafily find an excufe for it; it may well be faid, " that they muft have been more than men, had they acted lefs like men. Virtuous in the extreme that country would be deemed where the confcience of no individual, in the cool moments of reflection, could upbraid him with a heavier tranfgreffion, than applying to his own ufe a bit of iron that lay before him."

Chap. VII.

Paffage to Macao—Proceed to Canton and Embark for England—Anecdotes of Lee Boo—His Diftrefs and Death.—

THE Englifh were now once more on the way to all they held dear,

and

and having wiped away the tear of a friendly farewel, proceeded on their voyage with chearfulnefs; the firft two days the weather proved very tolerable; the wind varied from E. to S. E.

The principal perfon that will figure in the few following pages, is the highly valued Prince Lee Boo, a youth of the moft remarkable abilities, and in whofe hiftory every reader muft feel interefted. He had thrown himfelf freely into the protection of ftrangers, deferting his native country, his friends, his all, trufting to the honour of a handful of men of whofe exiftence he had but a few weeks before been utterly ignorant. For a few days he fuffered exceffively from fea ficknefs, in which fituation he could eat nothing, but was quite funk in his fpirits; this however wore off when he appeared to be eafy and contented. Captain Wilfon son now instructed him as to decency in his appearance, and desired he would dress as they did; he did so, but soon threw off the coat and jacket as insupportably cumbersome; but still retained the trowsers, as decency required, and would never afterwards part with them; indeed, as they began gradually to advance into a colder climate, he soon resumed the coat and jacket also. His notions of delicacy from what had been told him, and what he observed among the English, gradually increased, so that he would not so much as change any part of his dress unless when by himself. He continued to wash himself several times a day, and kept his clothes and every thing about his person very clean.

Sunday the 16th they kept a sort of thanksgiving to God for their deliverance,

ance, for which it may be safely said, they were all very grateful. They now also discovered a small leak in the vessel, which they found it impossible wholly to stop, they therefore employed two men constantly at the pump, which kept it under. Lee Boo was greatly discouraged at losing sight of land, a circumstance which had not previously occurred to him. From the 18th to the 25th of November, they had variable and disagreeable weather, heavy squalls, accompanied with thunder and lightning. On the 25th they came in sight of the Bashee Islands to the great joy of Lee Boo, who was happy once more, with at least a distant sight of terra firma. On the 26th they saw the island of Formosa, bearing N. E.; the 27th and 28th, the weather being favourable, they fell in with several Chinese

nese fishing vessels and small craft, and on the 29th anchored near the high land called Asses Ears. Having here engaged a pilot to conduct them to Macao, they arrived there next day. Lee Boo was greatly astonished at the size of the Portuguese ships in the Typa, calling out *clow, clow, muc clow!* that is, large, large, very large! The Governor paid Captain Wilson and his crew all manner of attention, and sent plenty of provisions of all kinds to the men on board the Oroolong, informing them at same time, that peace was now re-established in Europe. Captain Wilson, Lee Boo and the officers got lodgings appointed them on shore, except Mr Benger who took the command on board. An express was immediately forwarded to the Company's *supra-cargoes* at Canton, informing of their arrival and situation.

Mr M'Intyre an old acquaintance of Captain Wilson's paid them uncommon attention, and insisted on their lodging in his house. He had a Portuguese Gentleman in company with him, who invited them to his house on their way to Mr M'Intyre's, and that principally on Lee Boo's account, with whom he was greatly taken, and wished to introduce him to his family. His house was therefore the first into which Lee Boo entered, and his surprise on entering it cannot be easily described. The rooms, the furniture and ornaments, all severally crowded so many new objects on his mind at once, that he was perfectly lost in amazement; it was remarkable however, that amidst all his confusion, his behaviour was to the greatest degree easy and polite; and as he observed that he occasioned the same surprise in others that they did in him,

him, he very politely permitted them to examine his hands, described the tatooing, and appeared pleased with the attention paid him. On their way to Mr M'Intyre's, Lee Boo displayed his native benevolence very remarkably: observing the poor Tartar women, with their children tied to their backs, begging, he distributed all the oranges and other things he had about him among them.

When they reached Mr M'Intyre's it was late, so that the table was covered for supper, and the room elegantly illuminated; a new scene here burst upon him—the whole seemed to him a scene of magic. It is impossible to particularise every thing with which he was remarkably fascinated; a large mirror at the upper end of the room, rivetted his attention for a while; he saw his complete person, and supposed it to be somebody behind very like himself; he looked, laughed, and looked again, not knowing what to think. Indeed the mirror had a surprising effect on more than Lee Boo; the Englishmen had seen every one his neighbour's face, during all their distress, but nobody had seen his own; the hollow edged, long visag'd appearance they now made, to what they formerly recollected, cast rather a melancholy impression upon their minds. Next day Lee Boo spent mostly in examining Mr M'Intyre's house, in which he found abundance of new objects to surprise him. The other Gentlemen in the mean time went about purchasing such little commodities as they stood in need of, and every one brought in some little trinket with him for Lee Boo; among the rest, was a string of large glass beads, which almost distracted the poor Prince

Prince with surprise and joy. He conceived himself possessed of greater treasures than all the Pelew islands could afford; he run to Captain Wilson enraptured with his property, and begged that a small Chinese vessel might be hired to transport them to Pelew, and desired his father might be informed the *Englees* had carried him to a fine country, from whence he would soon send him other presents; at same time adding, that if the persons Captain Wilson should employ, faithfully and expeditiously executed their trust, he would reward them with two glass beads;——Happy state of innocence, where the utmost ambition can be so easily satisfied.

While at Macao, Lee Boo had frequent opportunities of seeing people of different nations, but soon gave a decided preference to the English, especially

ally the ladies. It has been already mentioned that there are no quadrupeds at the Pelew Islands, and that the Newfoundland dog left there, was the first of the species they had seen; they called him *sailor*, which name Lee Boo now applied to every quadruped he saw. Horses were his great favourites, he called them *clow sailor* or great sailor; he would often go to the stable and stroke their mane and neck, and soon ventured to mount them. He intreated Captain Wilson to send a horse to his uncle Raa Kook.

In a few days Captain Wilson received letters from Canton; the supra cargoes desiring him to draw for what money he needed, and ordering the men to be supplied with every necessary in abundance. The kindness shewn them on this occasion by all ranks at Macao and Canton, all the officers and

men

men fpeak of in the warmeft terms. Captain Wilfon, and his company took fhipping for Whampoa in the Walpole Captain Churchill, leaving Mr Benger to take care of the Oroolong, and difpofe of her. In a few days they arrived at Canton, having been kept in fpirits all the way by Lee Boo, whofe admiration at every new object and fenfible remarks, at fame time, furprifed all who had an opportunity of converfing him. He was greatly furprifed at the various difhes of meat, which he faw fucceffively fet before them, alledging, that his father, though a King, was happy to ferve himfelf with a few yams and cocoa-nuts; while here, the Gentlemen had a great many different dainties, and fervants attending them while they were eating. Obferving a man drunk, he faid he would not drink

drink spirits, as it made him unlike a man.

An instance of Lee Boo's strong attachment to those he knew, may be here mentioned, which will serve to point out this striking feature not in his character only, but in all the natives of Pelew. One day while sitting at a window which looked towards the sea, he observed a boat making towards shore, in which were Mr Benger and Mr M'Intyre, his joy was so great, that he did not take time to tell Captain Wilson or any other in the room the cause of his emotion, but springing from his seat, flew to the shore in a twinkling; immediately on their landing he shook hands with them so heartily and with such expressions of affection, as won their warmest regard. They had disposed of the schooner for
seven

seven hundred Spanish dollars, which was considered as a very good price.

Lee Boo became an universal favourite wherever he appeared; his agreeable good natured pleasant behaviour, made him acceptable in every company. He one day very much surprised a company of gentlemen with his dexterity in throwing the dart; a party was formed to have a trial of skill in the factory hall; they hung up a gauze cage and a bird painted in the middle; they stood at a good distance, and with much difficulty hit even the cage; when Lee Boo's turn came, he took up his spear very carelessly, and with the greatest ease, struck the little bird through the head. He had one day an opportunity of seeing some blue glass, which greatly delighted him; it was a colour he had not before seen; the gentleman in whose house it was,
made

made him a present of two jarrs of the same colour, which greatly delighted him; *Oh! were it possible he exclaimed, that my friends at Pelew could see them!*

As the time was now near at hand when the company's ships would sail for England, Captain Wilson laid before his people an account of the produce from the Oroolong, and other articles which had been sold; and giving to every one an equitable share, he addressed the whole company nearly as follows: " Gentlemen, the moment being now arrived when every one may to advantage follow his own inclination, I cannot part with you, without testifying my approbation at the spirited, the judicious, and the manly conduct you have preserved, amidst our trying difficulties; and be assured, that, on my arrival in England, I will represent you to the Company as entitled

titled to their particular regard, and I make no doubt but they will reward your toils." If mutual adverfity will reconcile the moft inveterate enemies, furely the fufferings of friends muft rivet a more forcible affection; and every one now prefent, appeared to feel the weight of the remark, for a tender concern was manifeft in every countenance. The conduct of Mr Wilfon had been fo mild and prudent, and his example fo animating, they ftill wifhed him for their commander: but it was not now a time for choice; neceffity and reafon pointed out their courfe. Lee Boo and the Captain came home in the Morfe, and the furgeon in the Lafcelles; while others embarked in different fhips.

What has already been mentioned concerning the amiable Lee Boo, has no doubt interefted every reader in his hiftory, history, in which we hope to be excused, if regard for a favourite subject, should lead us to be at times triflingly minute, as some readers may think.

The *Morse* was commanded by Captain Elliott, with whom Lee Boo made himself very happy; his spirit of enquiry concerning various objects which he saw, began now to be directed more concerning their utility than formerly; and he shewed no small anxiety to pick up as much knowledge as possible concerning such articles as would be useful at Pelew. His method of keeping his Journal is very noticeable; he had a string on which he cast a knot for every remarkable object he wished to imprint on his memory; these knots he examined daily, and by recollecting the circumstances which occasioned their being cast, he imprinted the transactions on his memory: the officers of the

the Morse humourously remarked when they saw him referring to his hempen tablet, that he was reading his Journal. He was not forgetful of the crew of the Oroolong, about whom he made frequent enquiries. Early in the voyage, he asked for a book, that with assistance he might learn the English alphabet, which was given him. At St Helena he was surprised at the sight of the soldiers, and the cannon on the fortifications; and four men of war arriving during his stay there, afforded new matter of astonishment. On being taken to see a school, he appeared so conscious of his own deficiency, that he begged he might learn like the boys. While here he had also opportunities of riding on horseback of which he was very fond; he galloped with great ease, and sat his horse very gracefully. Before the Morse sailed from St Helena, na, the Lafcelles arrived there, fo that he had an interview with his firft friend Mr Sharp; he was exceedingly happy with that Gentleman for whom he had the greateft regard. When the Morfe approached the Britifh channel, the number of fhips that paffed, confounded his Journal, and he was obliged to difcontinue his memorandums. But on landing at Portfmouth, the objects that met his view were fo ftupenduous and grand, he was involved in filent aftonifhment, afking no queftions whatever. The Captain proceeded to London, impatient to fee his family, and left Lee Boo under the protection of his brother; who, however, foon after fet off in a ftage-coach, with his innocent charge. Defcribing his journey, he faid he had been put into a little houfe, which horfes ran away with

with, and that though he went to sleep, he did not stop travelling.

On his arrival in London, he was not a little happy to meet with his Mentor, his new father, whom he was afraid he had lost. Being shewn his chamber, he could not conceive the use of the bed, it being a four post one and of course different from what he had seen on board: before he would repose himself, he jumped in and out of it several times, to admire its form, and intimating that here there was a house for every thing; it was all fine country, fine streets, fine coach, and house upon house up to the sky—for the huts at Pelew being only one story, he considered every floor here as a distinct house. Captain Wilson introduced Lee Boo to some of the East-India Directors, and to most of his friends; and at the same time shewed him the most conspicuous public buildings; but his prudent conductor kept him from stage and other exhibitions, left the heat of of the place might communicate the small-pox. He was sent to an acadedemy at Rotherhithe, where he was very assiduous in learning to read and write: and he soon became the favourite of all his schoolfellows, from his gentleness and affability. During the hours of recess, he amused his benefactor's family by mimicking such peculiarities as he observed in the boys at school. He said that when he returned to Pelew, he would keep an academy himself; and he imagined the great men of his country would think him very wise when he shewed them their letters. He always called his patron, Captain; but he would address Mrs Wilson no otherwise than his Mother, although he was to cold to the contrary;

contrary; conceiving it a tender expreffion. When he faw the young afking charity, he was highly offended, faying they ought to work; but the fupplication of the old and infirm met his natural benevolence—" Muft give poor old man; old man no able to work."

About this time he appeared to be about twenty years of age, middle fized; fo with having a moft expreffive countenance much fenfibility and good humour, that he inftantly prejudiced one in his favour; his eyes were fo ftrikingly expreffive that though he knew very little Englifh, his meaning was eafily underftood.

This quicknefs of manner and readinefs of apprehenfion were aftonifhing; a young lady with whom he was one day in company fat down to the harpfichord, in order to difcover how it affected him; to the mufic he paid little attention,

attention, but he was greatly interested to discover how the sounds were produced. He at same time sung a song in the Pelew style, but it was very harsh. He was naturally polite; one day at dinner, Mrs Wilson desired him to help her to some cherries, when Lee Boo very quickly proceeded to take them up with his fingers; she pleasantly hinted his error, when he immediately took up a spoon, at the same time his countenance was in a moment suffused with a blush.

Captain Wilson, one day, happening to rebuke his son for some trifling neglect in the presence of Lee Boo, the generous youth was not happy till he had joined their hands, which he did with the tears of sensibility streaming from his eyes. He preferred riding in a coach to every other conveyance, as it allowed people, he said, an opportunity ty of talking together: he was fond of going to church, becaufe he knew it was a religious duty, the *object* and final *end* being the fame both at Pelew and in England. He was prefent at Lunardi's aerial afcenfion; and remarked, that it was a ridiculous mode of travelling, as it could be done fo much eafier in a coach. He narrowly obferved all plants and fruit-trees, and faid he fhould take fome feeds of each to Pelew. Indeed, in all his purfuits, he never loft fight of what benefits they might might tend to in his country. But in the midft of his innocent refearches, juft as he was getting a quick knowledge of the language, he was taken ill of the dreaded fmall-pox: Dr Smith immediately attended him, who, in the firft ftage of the diforder predicted the fatal confequences which enfued. He cheerfully took the medicines that were adminiftered;

administered; and willingly dispensed with the sight of Mr Wilson, when he was told that he never had the disorder, and that it was infectious. In the midst of his illness, hearing that Mrs Wilson was confined to her chamber, he cried—" What, mother bad—Lee Boo get up to see her:" which he actually did. Mr Sharp, the surgeon of the Antelope, also attended him. Viewing himself in a glass just before his death, he turned his head away in disgust, at the appearance of his face, which was much swelled and disfigured. Getting worse, and sensible of his approaching fate, he fixed his eyes attentively on Mr Sharp, and said—" Good friend, when you go to my country, tell my father, that Lee Boo take much drink to make the small-pox go away, but he die—that Captain and Mother very kind—all English very good

good men—was much forry he could not tell Abba Thulle the great many fine things the Englifh got." He then enumerated all the prefents he had received, which he begged the furgeon to diftribute among his friends and the Rupacks. The dying difcourfe of this child of nature fo affected the man who attended him, that he could not help fobbing moft piteoufly, which Lee Boo obferving, afked—" Why fhould he cry fo, becaufe Lee Boo die?" Thinking Mrs Wilfon's illnefs arofe from his own, he would frequently cry out, fhe being only in an adjoining chamber, Lee Boo do well, mother:" The dreadful moment of feparation being now arrived, he told Mr Sharp he was going away;" and yielded his laft breath without apprehenfion, and with that native innocence and fimplicity which had marked his every action.

tion. The family, the servants, and those who knew him, could not withhold the tears of affectionate regard, when informed of the melancholy event;—The East-India Company ordered Lee Boo to be buried in Rotherhithe church-yard, with every possible mark of respect: all who knew him, with the pupils at the academy, attended the funeral; and the concourse was otherwise so great, that it might be supposed his good qualities had been publicly proclaimed, instead of being privately communicated. A tomb, with this inscription was soon after erected by the East-India Company:

To the Memory
Of Prince Lee Boo,
A Native of the Pelew or Palos Iſlands;
And ſon to Abba Thulle, Rupack or King
Of the Iſland Coorooraa*;
Who departed this Life on the 27th of December 1784,
Aged 20 Years,
This Stone is Inſcribed,
By the Honourable United Eaſt India Company,
As a Teſtimony of Eſteem
For the Humane and Kind Treatment
Afforded by his Father, to the Crew of their Ship,
The Antelope, CAPT. WILSON,
Which was wrecked off that Iſland
In the Night of the 9th of Auguſt 1783.

Stop, Reader, ſtop! Let *Nature* claim a Tear;
A Prince of *mine*, Lee Boo, lies bury'd here.

This amiable young prince, whoſe reſidence here was only five months, conformed himſelf to the Engliſh dreſs in every inſtance, except his hair, which he continued to wear after the faſhion

---

* Coorooraa is the proper name of the Iſland, of which Pelew is the capital town.

fashion of his own country. He was of a middling stature; and his countenance was so expressive, that it depicted the best qualities of a virtuous mind: his eyes were lively and intelligent; and his whole manner, gentle and interesting: he had the natural politeness of a gentleman, without the drudgery of study, or the observance of established forms of ceremony. After his death it was found that he had laid by all the seeds or stones of fruit he had eat after his arrival, with a view to plant them at Pelew.

When we reflect on the unhappy fate of poor Lee Boo, with which the reader is now acquainted, the mind ranges to the habitation of his father Abba Thulle, who on a cord had tied thirty knots, as a *memento* that his son would return in thirty moons or perhaps a few more; for which he was
willing

willing to make allowance. Those moons have long since performed their evolutions; the knots are untied; and yet no gladdening sail hovers round Pelew. Lee Boo is dead in reality; and though no more even in the tortured imagination of his expecting family, yet the sight of an European vessel, even at this distant period, would animate their hopes, and recal the fondness of past endearments. It will be a long time before the Ariel will reach this friendly, this hospitable shore; when the joy of the King to see a return of the English, will be so far overclouded by his parental disappointment: but his mind is too noble, open, and generous, to entertain for a moment, a suspicion that Captain Wilson could be guilty of inattention to Lee Boo, much less of baseness or ingratitude.

As a communication with the friendly

ly isles of Pelew is about to be renewed, there is every reason to hope that we shall yet receive a sequel to some of the preceding circumstances, with further particulars of a race of unenlightened people, whose sincerity, and strict adherence to the dictates of honour and religion, are at once a disgrace and a burlesque on the passions and pursuits of those who consider themselves as much nearer the standard of perfection.

F I N I S.

www.ingramcontent.com/pod-product-compliance
Lightning Source LLC
Chambersburg PA
CBHW032123230426
43672CB00009B/1836